EUREKA

EUREKA

HOW INVENTION HAPPENS

GAVIN WEIGHTMAN

YALE UNIVERSITY PRESS
NEW HAVEN AND LONDON

For information about this and other Yale University Press publications, please contact:
U.S. Office: sales.press@yale.edu www.yalebooks.com
Europe Office: sales@yaleup.co.uk www.yalebooks.co.uk

Typeset in Adobe Caslon Pro by IDSUK (DataConnection) Ltd
Printed in Great Britain by TJ International Ltd, Padstow, Cornwall

Library of Congress Cataloging-in-Publication Data

Weightman, Gavin.
Eureka : how invention happens / Gavin Weightman.
pages cm
ISBN 978-0-300-19208-7 (cloth : alkaline paper)
1. Inventions—History. 2. Discoveries in science—History. 3. Technological innovations—History. 4. Creative ability in technology—History. 5. Inventors—Biography. I. Title.
T49.5.W45 2015
600—dc23

2015007461

A catalogue record for this book is available from the British Library.

10 9 8 7 6 5 4 3 2 1

CONTENTS

INTRODUCTION

> Eureka moment n. an instant in which a scientific discovery is made or a breakthrough occurs . . .
>
> *Oxford English Dictionary*

All modern inventions have an ancient history. This is the central theme of *Eureka: How Invention Happens*, which is illustrated with an account of the long gestation of some familiar technologies that first appeared in the twentieth century and are now ubiquitous: the aeroplane, television, the mobile phone, the bar code and the personal computer. Each of these inventions was made possible by an accretion of earlier innovations and discoveries and it is their story which constitutes most of the history in this book. Here there are histories of many key technologies, from lithography to silk weaving to wireless, each of which played its part in the emergence of an entirely new invention for which it was not originally intended.

Starting with a 'eureka moment' when a new technology worked for the first time, however crudely, each chapter delves into the past to discover a breakthrough in scientific understanding that would lead eventually to the modern creation. The story comes full circle but stops short of the endless refinements which, for example, have produced the smartphone or digital broadcasting. The prehistory of an invention has a special fascination. Invariably the wisest and most knowledgeable of scientists and philosophers will have declared it an impossible dream right up to the time it is shown to be workable. Yet, once the first powered aeroplane has left the ground, the first image has flickered on to a television screen, the first grocery item has been scanned, the first personal computer has come to life and the first mobile phone calls are made, innovation is incredibly rapid. It is as if a seed has lain dormant for centuries until a favourable climate allows it to grow and bloom.

The point that inventions have to wait their time was made by Dr Vannevar Bush, Director of the US Office of Scientific Research and Development, in an article in *Atlantic Monthly* in 1945:

> Leibniz (1646–1716) invented a calculating machine which embodied most of the essential features of recent keyboard devices, but it could not then come into use. The economics of the situation were against it: the labor involved in constructing it, before the days of mass production exceeded the labor to be saved by its use, since all it could accomplish could be duplicated by sufficient use of pencil and paper. Moreover, it would have been subject to frequent break-down, so that it could not have been depended upon; for at that time and long after, complexity and unreliability were synonymous.
>
> Babbage, even with remarkably generous support for his time, could not produce his great arithmetical machine. His idea was sound enough, but construction and maintenance costs were then too heavy. Had a Pharaoh been given detailed and explicit designs of an automobile, and had he understood them completely, it would have taxed the resources of his kingdom to have fashioned the thousands of parts for a single car, and that car would have broken down on the first trip to Giza.

All inventions, however novel they seem when they first appear before the public, are reliant on earlier discoveries and innovations, so it can never be said that they are the work of a lone genius. However, it is striking that the inventor who makes the breakthrough is invariably outside the mainstream of existing industry and technology. Very often these inventors have been downright amateurs. The point was made by the once prominent American inventor Charles Jenkins in a little book he published in 1925. Had anyone noticed, he asked, 'the curious fact that a great laboratory, despite its inestimable contribution to science and engineering, has never yet brought forth a great, revolutionary invention which has subsequently started a new industry, like the telegraph, the telephone and telescope; motion picture, typecasting and talking machines; typewriter, bicycle and locomotive; automobile, flying machine and radio vision [an early form of television]. It has always been a poor man to first see these things, and as a rule the bigger the vision the poorer the man.'

This is not to put the amateur on a pedestal and to denigrate the scientist and the professional engineer for failing to imagine how a theoretical breakthrough might be turned into something practical. It is simply an observation that the 'eureka moment' so often falls to the outsider: not necessarily poor, but working with very limited resources and unencumbered by commitment to an existing technology.

Focusing on just five familiar technologies which first emerged in the twentieth century allows room in this narrative to conjure up something of the colour and complexity in the history of each. Working backwards from the 'eureka moment' offers an intriguing perspective: we find the bicycle an inspiration for the aeroplane, a talking automaton suggesting the telephone, early television dependent on discoveries made with a blowpipe and the microchip manufactured with a printing technique that dates from the nineteenth century.

Inevitably a history such as this has to deal on occasion with arguments over priority of invention, such as the patent dispute between Alexander Graham Bell and Elisha Gray: who first designed a working telephone? But such rivalries have never been a reliable guide to historical truth and they are of tangential interest. More significant by far is the contribution to innovation made by those who chose not to patent their discoveries, such as leading scientists Michael Faraday in England

and Joseph Henry in the United States. In fact it is not uncommon for inventors to give generous recognition to those whose discoveries made their own possible even though, to protect themselves financially, they might find themselves in disputes over patents. There is, for example, the touching story of the homage paid by Wilbur and Orville Wright to the German aviator Otto Lilienthal who inspired them with his daredevil glider flights. When they had succeeded with their invention of powered flight and had become wealthy and world famous, Wilbur paid his respects to Lilienthal's widow in Berlin and the brothers later sent her a handsome Christmas present.

The vast majority of inventions are not subject to any priority dispute, simply because nobody is interested in exploiting them commercially. It has been said that knowing *what* to invent is as important as being a creative technician, and one of the surprising discoveries of this history is that few people have the foresight to imagine that something novel – such as the telephone of the 1870s – will be popular. It was certainly not regarded as something that *had* to be invented to solve a contemporary problem. Notoriously, Sir William Preece, then chief electrician to the British Post Office, asked in 1879 by the Select Committee on Lighting by Electricity, which was concerned about interference from phones if he thought 'the telephone will be an instrument of the future that will be largely adopted by the public', replied that it might be in America but not in Britain. 'We have a superabundance of messenger boys, call boys etc., the Americans have not.'

This failure of imagination is found over and over again to be characteristic of those in established industries. Because inventors are so often trying to achieve what is conventionally regarded as hare-brained, those who are prominent in science or industry will distance themselves from the enterprise. There are examples in the history of scientists in the nineteenth century not wanting to be associated with those who were trying to build a powered aeroplane, as to lend the endeavour credence might brand them as cranks. The amateur and outsider has no such fears, which is why so much technology, such as the personal computer, starts out as a 'home-brew' and becomes mainstream only when it is shown to be practicable. It is not until then that industry adopts the invention and it becomes a necessity.

In reality, necessity is rarely the mother of invention. Who needed an aeroplane or a television before they were invented? What drives inventors, and has ruined any number of them, is the desire to show that something that everyone says is impossible could be made to work. Their success, if it comes, will be dependent on a great many earlier inventions as well as on the ingenuity of scientists and philosophers who themselves had no interest in how their discoveries might be turned into something of practical use. A classic example is the laser, which was developed without any thought that it would be ideal for reading bar codes.

The sense of 'eureka' as a moment of inspiration is very different from the concept of the 'eureka moment' – a point where there is a practical breakthrough in the quest to prove that the impossible is possible. There is nothing sudden about the eureka moments identified in this history: they are not simply 'good ideas', although they might involve a great deal of ingenuity. Invariably, the breakthrough comes after years of painstaking experimentation in which many different technologies are brought together. Yet those moments when, against all the odds, a new invention has been created are usually vividly exciting.

Few eureka moments can match the triumph of the Wright brothers who, throwing caution to the wind on the sand dunes of North Carolina in the winter of 1903, first got off the ground. It was a triumph which did not earn them instant fame and fortune but they knew they had achieved what many still believed would be impossible. Vital to their success was the enthusiastic help of a small community of fishermen and men from a Life-Saving Station who all along regarded Wilbur and Orville Wright as 'plumb crazy'.

CHAPTER 1

THE BIRD MEN

In the morning they raised the red flag to call the surfmen from the Life-Saving Station to help them haul the flyer into position on the windswept sand of Kitty Hawk. These lifeboatmen came with a sense of excitement for they had watched the evolution of this strange machine, which looked more like a giant winged insect than a bird. They were to be the only witnesses of an event rarely matched in the history of invention. For a long time the achievement was not believed; and the brothers themselves knew that it was only a beginning. Later, in the wooden shed they had built on the North Carolina coast, Orville Wright wrote in his diary:

> At just 12 o'clock Will started on the fourth and last trip. The machine started off with its ups and downs as it had before, but by the time he had gone over three or four hundred feet he had it under much better control, and was travelling on a fairly even course. It proceeded in this manner till it reached a small hummock out about 800 feet from the starting ways when it began its pitching again and suddenly darted to the ground. The front rudder frame was really badly broken up but the main frame suffered none at all. The distance over the ground was 852 feet in 59 seconds. . .

That day, 17 December 1903, Orville and his older brother Wilbur were the first in the world to fly a powered, piloted heavier-than-air machine. It was the fourth year they had camped out on the remote settlement of Kitty Hawk for several months, kept awake at night by the gales that constantly battered the coast, often frozen, short of food and, when a summer wind blew from the swamps, attacked by clouds of mosquitoes. This was their testing ground, carefully chosen for some of the very features that discomforted them. The wind gave them buoyancy in the air, the sand dunes offered soft landings, the shifting wind sculptured Kill Devil Hill, their glider launching pad. The soaring buzzards and eagles were an inspiration and models of aeronautical perfection.

The first year at Kitty Hawk they did not abandon their bicycle business in Dayton, Ohio until the summer season was over. Next year they brought in a mechanic, Charles E. Taylor, to look after the shop so that they could get away in summer. Their sister Katherine helped out as well. It was the craze for cycling which gave them the funds to experiment with flying machines and much of the equipment and expertise for building them. Between their first visit to the North Carolina coast in 1900 and the winter of 1903 they had relived the history of flight, reading all they could find, idolising some of those who had gone before. They worked methodically from kites to gliders and finally on their powered Flyer propelled by a lightweight petrol engine. The brothers always rested on Sundays as their father, Milton, was a bishop in the United American Church and would not have approved had they failed to observe the Sabbath.

On 14 December the brothers had tossed a coin to decide who should take the first ever powered flight. Wilbur won, but failed to control the flyer and it crashed. They had it repaired by the 17th when the conditions were not ideal for further tests: a bitter wind was gusting dangerously. Whenever a problem arose the brothers discussed it and came to a firm conclusion. On this day they decided they wanted to be home for Christmas and they would risk another attempt. As Wilbur had taken the first flight three days before, it was Orville's turn to take off but he was airborne for only 12 seconds. With their helpers they hauled the machine back and set it up for Wilbur to pilot. He managed about 12 seconds as well. On his next flight Orville rose about 14 feet off the ground before he was knocked sideways by a gust of wind. Finally, about midday, Wilbur climbed back on to the flyer, lying across the lower wing. When the propellers mounted at the rear of the flyer provided sufficient power, Wilbur launched himself into the stiff wind. This was the moment of triumph: a flight lasting one second short of a minute.

They took the machine back to their camp and were discussing the success of the flights when the wind turned the flyer over, so there could be no more attempts that day. The brothers had lunch and then set out on foot across the dunes to the weather station at Kitty Hawk about three miles from their camp. From here they were able to send a telegram home to their father in Dayton, Ohio. The cable, with Orville's name misspelled and the flight two seconds short, was wired first to Norfolk, Virginia on the weather station's government line. Addressed to Bishop M. Wright it read simply:

> Success four flights thursday morning all against twenty one mile wind started from level with engine power alone average speed through air thirty one miles longest 57 seconds inform Press home Christmas. Orevelle Wright.

There has been a long dispute about how this telegram got leaked to a local newspaper and how the information in the subsequent account of the flight was put together. It seems that Harry Moore, a 'rookie' reporter on the local *Virginian-Pilot*, had got to know John T. Daniels, one of the surfmen, after a chance meeting in a café in the town of Monteo. Moore was told about the crazy guys trying to fly out at Kitty Hawk. He asked to be kept informed about any success they might have and went to meet Wilbur and Orville, who did not know he was a journalist. Moore got a telegram from Daniels and another surfman saying the brothers had flown that day. Moore's editor, Keville Glennan, did not believe the story and spiked it. However, he changed his mind when the telegram to Milton Wright was leaked to another reporter, Ed Dean, when he made a routine call to the Kitty Hawk weather bureau. Everyone knew that attempts were being made in America and Europe to get a powered, manned, heavier-than-air flying machine off the ground. The ambition was widely dismissed as futile but this looked like a breakthrough. Editor and reporters put their heads together and concocted a story for which they had no more hard information than was contained in the cryptic telegram to the Bishop. Perhaps the wording of the telegram had started some Chinese whispers in the weather station for on 18 December the *Virginian-Pilot* ran the front-page headlines:

Flying Machine Soars Three Miles in Teeth of High Wind Over Sand Hills and Waves on Carolina Coast
No Balloon Attached to Aid it
Three Years of Hard, Secret Work by Two Ohio Brothers Crowned with Success
ACCOMPLISHED WHAT LANGLEY FAILED AT
With Man as Passenger Huge Machine Flies Like Bird Under Perfect Control
BOX KITE PRINCIPLE WITH TWO PROPELLERS

The description of the flight was wildly inaccurate but there was a kernel of truth in the story: 'While the United States government has been spending thousands of dollars in an effort to make practicable the ideas of Professor Langley of the Smithsonian Institute, Wilbur and Orville Wright, two brothers, natives of Dayton, O., have quietly, even secretly, perfected their invention, and put it to a successful test.' The professor was Samuel Pierpont Langley, a prominent American engineer whose government-backed steam-powered 'aerodromes' had plunged ignominiously into the Potomac River at the moment of launching, for reasons which will become clear.

The finale of the *Virginian-Pilot's* fantasy report on the Kitty Hawk flight had Wilbur selecting a suitable place to land where his flyer 'settled, like some big bird, in the chosen spot. "Eureka" he cried like the alchemist of old.' The flights that day certainly represented the 'eureka moment' for the Wright brothers, but it would have been quite out of character for either of them to make such an exclamation. They had a remarkable formality about them, dressing in wing collars and smart suits and looking for all the world like two comedians from a silent movie. But they were excited. Though the best flight was short of a minute they knew they had made the breakthrough.

After the first newspaper reports were published, all repeating the flight of fancy concocted by the *Virginian-Pilot*, there were one or two enquiries back in Dayton, Ohio where Bishop Milton was handling publicity. He made a brief statement expressing his great pride in the achievement of his two youngest sons:

> Wilbur is 36, Orville 32, and they are as inseparable as twins. For several years they have read up on aeronautics as a physician would read his books, and they have studied, discussed, and experimented together. Natural workmen, they have invented, constructed and operated their gliders, and finally their 'Wright Flyer' jointly, all at their personal expense. About equal credit is due each [written on 22 December 1903].

In the end, not much notice was taken of this historic flight. Many did not believe it had taken place. Others were unimpressed with a flight that had lasted only 59 seconds. What use was that? The brothers abandoned the Kitty Hawk testing ground and decided to trial new and better models nearer to home. Until the success of the flights on 17 December they had regarded their experimental gliding and attempts at powered flight as a pleasurable pastime. They now reasoned that to take it further and to create a Wright Flyer that had some practical use, they would have to invest in it much more heavily. As Wilbur put it, they had 'reached a fork in the road'. Either they would go back to their bicycle business, which made them a decent living, and treat flying as a hobby, or they could make aeronautics their main business.

The brothers were aware that the desire to emulate the majestic ability of the great soaring birds to stay airborne almost effortlessly had doomed countless pioneers of flight to disappointment and not a few to death. They were not interested in getting airborne by ballooning, nor did they regard the airship as a rival to their Flyer. For true flight, men would somehow have to get airborne on wings. Few believed it possible, with or without some form of power: the shattered bodies of what an aviation historian has called the 'tower jumpers' were grim evidence of the futility of man's ambition to fly. However, there was one very early student of aeronautics who got so close to finding the solution that some accord him, rather than the Wright Brothers, the title of inventor of the aeroplane though he lived at a time when the technology to turn theory into practice did not exist.

* * *

Some time in 1808 news reached Sir George Cayley of Brompton Hall near Scarborough in Yorkshire that an ingenious Viennese clockmaker called Jacob Degen had confounded the scientific wisdom of the day and had actually taken to the air under the power of his own arms in a machine called an ornithopter. The first reports were of Degen flying in a high-ceilinged building in Vienna to the delight of his friends and onlookers. He had strapped to each arm lightweight wings, which he flapped. As these did not provide enough power to get him off the ground he devised a hoist using a counterweight to get airborne and then bounced around rather like a bird on string. After a few trial runs Degen took his ornithopter out into the open where it caused a sensation. Illustrations were circulated of the pioneer aviator in the air and reports assured sceptical readers that Degen had indeed flown like a bird.

This account of manned flight was encouraging for Sir George, who had turned his richly inventive mind to the problem of flight a number of years earlier and had conducted a good deal of research into the science of aerodynamics. Some aspects of this subject were regarded as worthy of study by the scientific establishment of the day: for example, ballistics, research into the most favourable shape for a bullet, or the relative effectiveness of different designs of windmill sails. But to take seriously the notion that a winged man might fly was to invite ridicule.

As a young man, Cayley had witnessed the craze for ballooning, which was popular in the early nineteenth century. However, this form of aerial travel had an obvious limitation, which was put succinctly by Dr Johnson in a letter written in 1784: 'In amusement, mere amusement, I am afraid it must end, for I do not find that its course can be directed so that it should serve any useful purposes of communication.' It was a view Cayley shared. His interest in aviation had been inspired not by the balloonists but by a little toy ornithopter which he had made to a well-established design and which rose to the ceiling when its feather propellers were spun by a tug of a thread. Unlike a balloon, this was a heavier-than-air machine and its flight more closely mimicked that of a bird.

Believing that Degen the clockmaker had shown that man could fly and that his own endeavours could no longer be dismissed as pointless, Cayley submitted three pieces to the obscure and short-lived *Journal of Natural Philosophy, Chemistry and the Arts,* setting out his aerodynamic theories. The first article, dated 6 September 1809, began: 'Sir, I observed in your Journal for last month that a watchmaker at Vienna, of the name of Degen, has succeeded in raising himself in the air by mechanical means. . . . It appears to me, and I am more confirmed by the success of the ingenious Mr Degen, that nothing more is necessary, in order to bring the following principles into common practical use, than the endeavours of skilful artificers, who may vary the means of execution, till those most convenient are attained.'

Noting that the idea of flying with artificial wings had for long been ridiculed, Cayley was confident that, through his own endeavours, he might be 'expediting the attainment of an object, that will in time be found of great importance to mankind: so much so, that a new era in society will commence, from the moment that aerial navigation is familiarly realised'. After reading Cayley's articles, a Fellow of the Royal

Society, Sir Anthony Carlisle, wrote a private letter of support in which he explained that his own interest in aviation had been suppressed: 'The Wise, the Prudent and the Cunning Classes of Philosophers are too wary to commit themselves on subjects not backed by the cry of the multitude and had I been able to write such papers as you have printed, I should have been ridiculed and abused to my irreparable injury.'

As it turned out, Degen was a fraud. He had his comeuppance when, in 1812, he charged spectators in Paris a hefty fee to watch him flapping his wings as he leapt into the air. Omitted from the programme notes and the illustrations advertising the demonstrations was the fact that Degen was supported in the air by a hot air balloon without which he would have been firmly rooted to the ground. Reports in *Le Journal de Paris* described his denouement at the hands of angry spectators. It seems he was badly beaten and was subsequently lampooned in song and on stage, and nicknamed *Vol-au-Vent* after the hollow puff pastry case that was a favourite of French cuisine.

The favourable reference to Degen was cut by unknown censors from later reprints of Cayley's 1809 article, presumably to save him the embarrassment of being duped. However, Cayley himself does not appear to have been discouraged by the deception and held to his conviction that manned flight was possible, in doing so staking his claim as an inventor proverbially ahead of his time. Here was yet another example, it would appear, of a golden rule of invention: someone conceives of a workable, theoretical model long before the technologies that will make it practical have evolved. In the history of aviation, Sir George Cayley was the inventive genius who achieved that. He did not quite arrive at the solution the Wright brothers discovered: he preferred paddles to propellers to power his imaginary aircraft. It should remembered, however, that when he was designing his first aircraft the earliest steam-driven ships used paddles and the screw propeller did not become widely used at sea until the mid-nineteenth century. And, although he was fully aware of the problem, Cayley lacked the practical experience in flying a full-sized glider which would have enabled him to discover a means of stabilising an aeroplane in mid-air. But it is now generally agreed by generations of aeronautical historians that Cayley grasped the basics of heavier-than-air flight before anyone else and that at least some of his experimentation was of value to pioneer aviators at the end of the nineteenth century.

Born into the Yorkshire squirearchy in 1773, Cayley had a rather unusual education directed principally by his mother Isabella Seton whose Scottish family motto was ' Ever Gay'. She had a lively interest in political and scientific debate and sought a suitable education for a son who had early on shown an interest in mechanics. George was sent first to a school in York but he became ill there so his mother took him out and arranged for him to be tutored privately. At the age of eighteen he moved to Nottingham to live with George Walker, a Nonconformist minister, mathematician and Fellow of the Royal Society. Walker had one daughter, Sarah, who studied with his lodgers, and George fell in love with her. She was very bright and attractive and two years older than George but his mother disapproved of the attachment and had him moved to another tutor, George Cadogan Morgan, who lived in Southgate just to the north of London and lectured at Hackney College in mechanics and electricity.

Cayley was immersed in a world of learning: his tutors were on friendly terms with all of the great thinkers and philosophers of the day, including Joseph Priestley. However, his time there was cut short when his father died and he succeeded to the baronetcy when he was still just nineteen. Three years later he married Sarah Walker and embarked on a domestic life which was, by all accounts, tempestuous. In their affectionate biography of Cayley, *The Life of a Genius*, Gerard Fairlie and Elizabeth Cayley give a colourful account of this extraordinary woman who alarmed the family, their friends and tenants alike:

> They were shocked by her tendency to lose her temper, frequently and alarmingly, no matter who was present. She had other unconventional ways, extremely unconventional in those days, which made it impossible for her to win the affection of the people around her early in her marriage. Ladies just did not smoke. Sarah smoked, and

> smoked a pipe, with obvious enjoyment even in public. No lady rode astride: Sarah did. She preferred masculine sports. She was a fearless and headstrong rider to hounds and had been taught, and permitted, by her father to use his shotgun.

This caricature is perhaps unfair, for Sir George never lost his affection for Sarah who, despite her supposed masculinity, had ten children, seven girls and three boys, two of whom died in childhood of measles. As a mother she was apparently neglectful and difficult, and much of the childcare was taken up by George's cousin, known as Miss Phil (short for Philadelphia) who clearly doted on him and remained his close friend and confidante all his life. Towards the end of her life Sarah became deranged.

As a baronet and the local squire, Cayley clearly had a paternal concern for those who were not born into such a privileged position as himself. He turned his inventive mind to many problems, from the drainage of his estate to railways, for which he designed an automatic signalling system and a safety brake. For a millworker whose hand had been crushed he worked with his resident engineer, Thomas Vick, to fashion an artificial hand that worked. Cayley's workshop in an outbuilding of Brompton Hall on his estate was always full of devices and drawings.

Although his first accounts of what he called ' the principles of aerial navigation' were not published until 1809, he had been intrigued by the possibility of flight since boyhood. Much of the archive he left has been unearthed haphazardly over the years and it is only recently that some sketches he made at school of flying machines have been discovered. But the strangest and most haunting object, which some believe encapsulates the essential principles of the aeroplane, came to light long after his death.

In 1925 an elderly lady, who said her mother was from the Cayley family, sold a collection of small silver pieces to a watchmaker and silversmith in Scarborough. Among them was a disk on one side of which was engraved what is clearly a flying machine and on the other a diagrammatic representation of the forces acting on an object in flight.

Below the flying machine are engraved the letters 'G C 1799' and below that the letter 'R'. Though there are literally thousands of aeroplane designs that pre-date this engraving, including those of Leonardo da Vinci, this is judged to be the first truly 'modern' depiction of a fixed-wing flying machine.

It does not look like a modern aeroplane, however. The pilot is enclosed in what resembles an open boat and the tail fins are cruciform. The diagrammatic representation of the forces acting on a wing is not readily interpretable by someone with no understanding of aerodynamics. But there is a consensus that by 1799 Cayley had understood a fundamental principle that was to make heavier-than-air flight possible. It was some time after the disk was bought that its significance was recognised and it was presented to the Science Museum in London, where it remains today.

In his 1809 article Cayley stated clearly the discovery encapsulated on the disk:

> The whole problem is confined within these limits, *viz*–to make a surface support a given weight by the application of power to the resistance of air.

For Cayley, as for others before and after him, the bird which soared and glided through the air with no appreciable effort was the model for his plane, its movement and balance achieved by the action of air on its wings the shape of which was crucial in providing lift and drag. For propulsion the bird could use a beat of its wings or the wind or rising warm air currents against the draw of gravity. If a man were simply to glide like a bird he would not need any means of propulsion, provided he started from a high enough point above ground. But to fly continuously he would need a source of power.

In 1804 Cayley built a simple, hand-launched model glider and in his notebook gave a description of its launch in the hills of the Yorkshire Dales. This, it is generally agreed, is the first ever description of a true, fixed-wing aeroplane flight:

> It was very pretty to see it sail down a steep hill, and it gave the idea that a larger instrument would be a better and safer conveyance down the Alps than even the surefooted mule, let him meditate his track ever so intensely. The least inclination of the tail towards the right or left made it shape its course like a ship by the rudder.

Cayley's glider was made from what he called a 'common paper kite' fastened to a wooden rod to form the wings set at a slight upward angle, with a tail shaped like the feathers of a dart. Its design was based on calculations which related the speed of flight to the weight of the glider (3.82 ounces) and the area of the wings. He compared its design to that of the crow, a bird he had studied in great detail, paying special attention to the structure and angle of its wings in flight. A critical factor was the centre of gravity of the glider, which Cayley varied by moving a weight along the central rod.

Up to 1818 Cayley conducted a great many experiments in his workshop aimed at understanding how the resistance of air acted on wings of difference shape and tilt. He created a whirling arm mechanism that is the forerunner of the wind tunnel used to study airflow over wings. He designed motors he hoped would give a better weight-to-power ratio than the steam engine. One was a form of internal combustion engine in which the power came not from ignited gas or petrol fumes but gunpowder. The other was powered by hot air. Although he recognised the limitation of balloons, which could not be steered, he designed an airship very like those that eventually flew. These were cigar shaped and had the buoyancy to carry steam engines which drove propellers. He anticipated, quite correctly, that these airships would provide a practical form of air transport before the fixed-wing aeroplane, the design of which presented much greater problems. However, after 1818 Cayley turned his mind to other concerns and was not inspired to return to the subject of aerial navigation until he learned of a bold venture that promised to make powered flight a reality.

Whilst he had been taken in by the story of the flying Viennese watchmaker Degen, Cayley was not fooled by the widely promoted invention of two lacemakers from Chard in Somerset, England. If the publicity was to be believed, they had built an aeroplane which could carry not only a pilot but passengers and would shortly be transporting tourists and businessmen all over the world at unheard-of speeds. A patent had been granted in 1843 and funding sought by the Aerial Transit Company which was being advertised with wonderfully futuristic posters showing a flying machine just after its launch from a platform setting off from a variety of exotic locations.

Beneath its huge wings was slung a boat shaped body with three wheels, presumably its landing gear, and it was powered by two whirring propellers mounted behind the undercarriage. It was the brainchild of William Henson, working with a number of associates, who included a fellow lacemaker John Stringfellow and an instrument maker, John Chapman. Though nobody had ever witnessed this wonderful creation in action, Henson was given credence and glowing publicity of a kind that had never before been accorded a pioneer aviator.

Under the heading 'The Aerial Steam Carriage' *The Times* wrote on 30 March 1843:

> When it was announced some time ago, by the ordinary monthly lists of patents granted, that Mr Henson had invented a machine capable of conveying despatches and passengers through the air, the general impression was that some moody and enthusiastic projector was about to exhibit the produce of his day dreams. Our readers may, therefore, be somewhat surprised to learn that this is in truth no visionary scheme, but a design of very scientific conception, carefully and perseveringly wrought out. It would perhaps be too much to affirm (what yet we cannot deny) that the machine in its present state will certainly succeed; but the least which can be said is, that the inventor has most skilfully removed the difficulties which have hitherto debarred us the possession of the long coveted faculty of flight, and made its eventual, perhaps early, attainment a matter of little less than certainty.

The article, which continues in this vein for several more paragraphs, is unsigned but might well have been written by Henson's promoter, Frederick Marriott, a journalist and publicity agent based in Chard, centre of the lace-making industry and Henson and Stringfellow's home town. There is precious little detail about the supposed aerodynamics of the aircraft, and what there is startled Sir George Cayley when he read about Henson's proposal. In particular, the single wing span of 150 feet seemed outlandish. However, the novelty which *The Times* imagined would get Henson's aerial steam carriage off the ground was the miniaturised engine. Until that time the weight to power ratio of steam engines had made them quite impractical for flight.

Although the romantic depictions of the Henson aeroplane captured the imagination of newspapers and the public, they did not attract any investors. This presented Henson and Stringfellow with a huge problem, for they had not yet begun to build their aerial steam carriage. In fact they did not even have a model. The full-scale aircraft never was built and Henson despaired, married and emigrated to the United States. Stringfellow lost interest for a while but did eventually produce a model steam-driven plane which perhaps achieved a very short flight. However, the abject failure of the Henson–Stringfellow enterprise is still regarded as significant in the history of aeroplane design.

First, it revived Cayley's interest in aeronautics and he designed, and probably had built, two full-scale gliders. The first, which he called The Boy-Carrier, he designed in 1849 with the intention that it should be powered by an engine of some kind, which would work flappers to drive it forward. It was tested only as a glider, however. 'The balance and steerage was ascertained, and a boy of about ten years of age was floated off the ground for several yards on descending a hill [that is, gliding down], and also for about the same space by some persons pulling the apparatus against a very slight breeze by a rope.'

Commenting on the huge intended wingspan of the Henson machine, Cayley suggested it would be better to achieve the same wing area by creating a triplane: the wings would be shorter, with one above the other. Stringfellow incorporated this idea into a model he built in the 1860s. Again, the model was widely publicised without achieving much in the way of flight. But what it did achieve was to suggest a prototype for future innovators. So too did the original aerial steam carriage, the illustrations of which anticipated the shape of aircraft that did fly: the fixed wing, the propellers, the cabin and the undercarriage. However, there is no doubt that had this aeroplane been built and an attempt made to fly it, it would have ended up a heap of broken timber and twisted metal. Its resemblance to more modern aircraft is superficial.

A great deal more had to be learned about the forces that Cayley had identified, the secrets of which enabled gliding birds to lift off effortlessly into the wind. Attempts were made to understand the nature of airflow and air resistance, but the physics of aerodynamics was extremely complex and it became clear to would-be aviators that the only way to discover how to fly was to take the huge risk of experimenting in mid-air. There were a great many prepared to try it but none more enthusiastically and methodically than a flamboyant German with a taste for public spectacle.

* * *

When they knew the 'Flying Man' was going to launch himself on a Sunday afternoon from his purpose-built *Fliegerberg* (Flyer's Hill) a crowd of Berliners would gather, eager to see for themselves the spectacle which was regarded by the public and scientists alike as a daring and foolhardy stunt. This was the summer of 1894 shortly after a special commission of German scientists had concluded finally that it was impossible for a man to fly. Yet, there, on top of the specially created *Fliegerberg*, was a middle aged man wearing a red sweater and knee breeches, his arms strapped into two huge wings, preparing to leap off into thin air in defiance of the predictions of the country's most eminent scientists. The Flying Man, Otto Lilienthal, was not at all troubled: he had been doing this for several years and he was still in one piece. And so with a short run he launched himself from the hut he had built on top of the *Fliegerberg* and began a glide that would carry him, legs dangling, down the slope of the hill with the distance and angle of his descent governed by the strength of the breeze.

It was a theatrical performance and intended as such, for the Flying Man was keen to promote gliding as a sport to rival the very recent craze for cycling. He had taken advantage of the newly devised high-speed cameras to get photographers to snap him in mid-air and he had attracted a good deal of attention in the newspapers. The American magazine *McClure's* ran a picture feature on him in September 1894 illustrating flights with a variety of gliders. Under the cross-head 'The Bird's Wing Lilienthal's Model' was a brief account of this pioneer aeronaut's history: 'For more than twenty years, Herr Lilienthal, with his brother's aid, and in the intervals of more serious occupations, has been studying the subject of aerial navigation. He has taken the flying bird as his teacher.' Otto, the Flying Man, was 'an accomplished mathematician and a close observer of nature; and is, besides, endowed in large measure with that poetic instinct which nearly always constitutes one side of even the most practical German character.' A writer for the *Dresdner Nachrichten* (Dresden News) described graphically on 17 November 1894 a future full of 'flying people' à la Lilienthal. For example: cobblers' delivery boys who fly with glowing cigarettes clenched in their teeth from one customer to the next, or illicit lovers, whose wings would 'come in handy in the event of unpleasant surprises'.

Otto's brother, Gustav, a year younger, did not attempt to fly and did not always share Otto's obsession with gliding. For much of their lives, Otto was the breadwinner, the more successful industrialist, and neither Gustav nor Otto's wife, Agnes, wanted him to go on risking his life.

There is no doubt that, with his studies of bird flight, the structure of birds' wings and his own experience of gliding, Otto Lilienthal was to be hugely influential in the decade leading up to the eventual achievement of manned flight. But there is something of a mystery about how he and his brother arrived at their approach to the problem of heavier-than-air aviation. In 1889, Otto published a book entitled *Der Vogelflug als Grundlage der Fliegekunst* (Bird flight as the Basis of Aviation) in which he set out all the experiments he and Gustav had made in an attempt to study and formulate the functioning of a bird's wing which appeared, in miraculous ways, to allow for buoyancy and speed with a minimal amount of effort.

The book is written as if they started from scratch with absolutely no notion of previous attempts to address the same problem. It resembles a fairy tale, with a touch of that Germanic poetry that *McClure's* detected. When Otto's book was first published in an English translation in 1911, Gustav gave an account of how they had become interested in flight as children. There was the story of Count Zambeccari who perished in 1812 when he ran into a storm in the balloon he was attempting to steer with paddles. And there was the fable of the willow wren and the stork. Gustav wrote: 'The small willow wren happens to meet the stork, and complains of fatigue; the latter, in his generosity, offers him a seat on his back, and during the ensuing conversation the stork explains the method by which he sails without effort or wing-beats, and how he planes down in a straight line from a great altitude to a distant meadow.' Storks nested on the roofs of the barns in the countryside around their home town of Anklam in northern Germany and the brothers were able to study them, noting how, when startled, the huge birds struggled to take off, always running into the wind. Once airborne, their flight seemed almost effortless.

Whether or not the brothers would have pursued their interest in flight if they had enjoyed an untroubled childhood it is impossible to say. But it was tragedy that, by chance, opened up possibilities that might otherwise not have presented themselves. Four of their younger siblings died, three sisters aged from a few months to four years old and a brother at less than one year. One sister had survived and was five years old when the last of the Lilienthal children, Anna, was born in 1861. This prompted their father, Gustav, to make up his mind that he would join the exodus of Pomeranians who at that time were emigrating in large numbers to the United States.

Gustav was a physically large man who had a gift for mechanics and mathematics but who had been persuaded by his father to go into the drapery trade. He had to borrow the money from his father to set up in business and was perpetually in debt. Nevertheless, he was able to court and marry an exceptional woman who at one time had begun a career as an opera singer. Caroline Pohle was also from Ankram but had gone to study in Berlin under the patronage of a wealthy aunt. When the aunt died and she could no longer pay her fees, Caroline returned to Anklam and fell in love with Gustav. They were married in 1845. It was a turbulent time in Europe, with many uprisings, and Gustav's business struggled until 1854 when he was declared bankrupt. The Lilienthals had to move to a cheaper house and finally to seek their fortune in America. Before they could leave Gustav died of a virulent form of pulmonary tuberculosis. Six months later the infant Anna died.

Caroline wrote in her diary three years later: 'My greatest good fortune lies in my three children, Otto, Gustav and Marie who I am endeavouring to bring up to be good people.' She somehow took strength from her misfortune, did not remarry, and set about making her children's lives as enjoyable and interesting as possible. According to Gustav, she encouraged the inventiveness of her sons. 'Our mother fostered in every way our mechanical proclivities, and never refused us the means to purchase the requisite materials for our experiments, however hard it may have been for her at times. Well do I remember submitting to her our plans for our first flying machine, to the construction of which she readily consented.'

As teenagers Otto and Gustav made some crude attempts at flying, only to discover what so many had before them: strapping on a pair of wings and attempting to take off was quite beyond the capability of man. These youthful experiments came to an end when their mother arranged for their education. Otto went off to Berlin where he worked for a year in an engineering factory making parts for the rapidly expanding railway network as a prelude to his enrolment at the Berlin Industrial Academy. Here he received a sound education, excelling at mathematics and engineering. Gustav had a more modest schooling and later joined his brother in Berlin. During holidays they began their scientific experiments to test the lifting potential of wings of different shapes and construction. On an uncle's estate near Anklam they built a kind of gallows attached to a barn wall from which was suspended on a rope a treadle-operated wing mechanism. With the assistance of their sister Marie they noted measurements of lift as they strained to get their artificial wings to flap. They also created a mechanical wing mechanism, which gave them what they believed were accurate readings. This was in the late 1860s and it provided data that would not be published or put into practical use for nearly thirty years.

That was not the end of the brothers' youthful collaboration on flight, but there was little time or money for experimentation as they struggled to make a living. In 1870 war between France and Prussia drew them apart, Otto enlisting and Gustav, excused military service on health grounds, staying in Berlin. Gustav had worked as a bricklayer before enrolling at the Berlin Building Academy but the war put an end to his studies there. He began work as an architect and when Otto returned to Berlin in 1871 they continued to experiment with model planes, including one the size of a stork which had flapping wings driven by a miniature steam engine. This was a complete failure, breaking up on its maiden flight.

In 1872 their mother Caroline planned to sell the family home in Anklam and move to Berlin. Before she had sold off most of the furniture she caught a chill which developed into pneumonia, and she died at the age of forty-six. Otto, Gustav and their sister Marie moved in with their grandmother to form a new family household. Gustav had wanted to travel, threatening sometimes to emigrate to America as he could find no fulfilling work in Berlin. In 1873 he got the chance to leave Germany when he found work as an architect in Prague. A short time later, in the same year, he was invited to London by the English architect William Crossland, who was engaged in some major projects including the building of Royal Holloway College. How this invitation came about is not known, but it is tempting to speculate that it had a far-reaching influence on the brothers' experiments with flight.

While he was in London, Gustav was introduced by Crossland to the Aeronautical Society of Great Britain. He gave a talk in which he described the experiments he and Otto had conducted. He was very impressed with the British interest in heavier-than-air flight, which was dismissed as fanciful in Germany, and he enrolled Otto and himself as members of the Society. In a biography by Manuela Runge and Bernd Lukasch, *Erfinderleben: Die Brüder Otto Und Gustav Lilienthal* (translated as Inventor Brothers: The Lives and Times of Otto and Gustav Lilienthal) there is the following speculation:

> Gustav must surely have learnt about one kindred spirit who had also been alone in the field. . . . The works, books, drawings and models of Cayley kept by the Society would have been of immense interest to Gustav, and he must surely have taken the chance to explore and to absorb them, and draw inspiration from them. When Gustav returned to Berlin in May 1874, he was bubbling over with ideas, both as an architect and as a pioneer of flight.

In terms of the science of aerodynamics and the optimal shape of a wing, Cayley clearly anticipated the later discoveries of the Lilienthals and yet he is not mentioned in Otto's book *Birdflight*. Neither is the English inventor Horatio Phillips who, in 1884, built a wind tunnel powered by injected steam to test the lifting potential of different aerofoil (wing) shapes. When Otto arrived at more or less the same design he found it already patented by Phillips. One of the biographers of the Lilienthals, Bernd Lukasch, suggests that Otto wanted to write a popular book to promote gliding rather than a technical treatise. Whatever the reason, the impression the book gives is that he and Gustav thought it all out themselves.

Perhaps the chief influence of the discoveries of Cayley and other pioneers of aviation on the Lilienthal brothers was the fact that they found in them support for their view that human flight was possible and that the key to it was the study of aerodynamics. There was far less interest in heavier-than-air flight in Germany than in France or Britain, and when he began to put theory into practice Otto clearly felt himself to be alone in a deeply sceptical world. He was confident in his own scientific approach to the problem of flight and there is no suggestion that in designing his gliders he copied those of Cayley or any other pioneer. What is so remarkable about Otto Lilienthal is that he really was the first to design, build and fly a glider – today we would think of it as more like a hang-glider – and to demonstrate to the world in a theatrical fashion that human flight was possible.

Between the youthful experimentation of the Lilienthals and Otto's first tentative flights there were many years when no progress was made at all. Gustav tried and failed to make a fortune with many enterprises and for a while lived in Australia, taking his sister Marie with him. She met a farmer from New Zealand on the boat, then married him and went to live in New Zealand. Gustav returned to Germany a single man. Meanwhile Otto established a business making compact 'safety' steam engines for small workshops, his fortune rising and falling with the vagaries of the German economy. He indulged his love of the theatre, performing and directing in Berlin.

In 1889 Otto published *Birdflight* and two years later began his series of gliding experiments in the garden of his home. At first he just made a few hops to get the feel of being airborne but soon he looked for more open ground where he could attempt longer flights. It was at this point, when the dream of flying was about to become a reality, that Gustav began to take the view that what Otto was up to was a waste of time and money. He told people he thought it was all nonsense. And it was true that Otto was spending more and more money on his flying experiments.

Otto himself believed that if he could create enough enthusiasm for gliding it would become a craze like cycling – a sport he enjoyed – and that he could build a successful business selling his patented designs, or 'normal gliders' as he called them. He would promote the sport from his specially built *Fliegerberg* and when public interest began to grow he would begin to manufacture and sell gliders. He did sell a few: how many we do not know. But the sport did not catch on. Most of those who took up the challenge of flying one of his gliders did not want a second go: it was too difficult and too frightening.

Between his first short hops in 1891 and his longest glides in 1896 it was estimated that Otto made 2,000 flights. He was not reckless. The gliders were designed so that he could slip out of them if he felt he was being dragged dangerously. He progressed in stages, testing his ability to control the glider in the air by shifting the weight of his body. He believed if he could get up high enough he might be able to manoeuvre the glider so that he could turn in the air, describing a circle.

The *Fliegerberg*, purpose-built on the spoil heap of a brickworks, was his showground but not where he would put himself to the test. He could get longer glides in the Rhinow Hills near Stolin. Here there might just be a few friends or members of his family to watch, and his devoted mechanic Paul Beylich, the son of a blacksmith who lived near the *Fliegerberg*. On occasion there were visitors from abroad. One of these was the American physicist and photographer Robert W. Wood, who watched Otto flying in the summer of 1896 and wrote a wonderfully evocative piece for the *Boston Evening Transcript* which captures the thrill of a Lilienthal flight:

> A brisk wind was blowing, and the storks were sailing over the fields on each side of the road in search for food for their young on the chimney-tops. . . . We had a hurried lunch in the little inn at Rhinow, where his arrival always causes a hum of excitement among the peasants; the flying machine was brought out of the barn, and loaded on the wagon, and we drove away to the mountains. . .
>
> The machine was laid out on the grass and put together . . . in the bright sunshine with its twenty-four square yards of snow-white cloth spread before you, you felt as if the flying age was really commencing. Here was a flying machine, not constructed by a crank . . . but by an engineer of ability. . . . We carried it to the top of the hill, and Lilienthal took his place in the frame, lifting the machine from the ground. He was dressed in a flannel shirt and knickerbockers, the knees of which were thickly padded to lessen the shock in case of a too rapid descent. . .

> I took my place considerably below him by my camera, and waited anxiously for the start. . . . Presently the breeze freshened a little; he took three rapid steps forward and was instantly lifted from the ground, sailing off nearly horizontally from the summit. He went over my head at a terrific pace, at an elevation of about fifty feet, the wind playing wild tunes on the tense cordage of the machine, and was past me before I had time to train the camera on him.
>
> Suddenly he swerved to the left, somewhat obliquely to the wind. . . . It happened so quickly and I was so excited at the moment that I did not quite grasp exactly what happened, but the apparatus tipped sideways as if a sudden gust had got under the left wing. For a moment I could see the top of the aeroplane, and then with a powerful throw of his legs he brought the machine once more on an even keel, and sailed away below me across the fields at the bottom, kicking at the tops of the haycocks as he passed over them. When within a foot of the ground he threw his legs forward allowing the wind to strike under the wings, and he dropped lightly to the earth. I ran after him and found him quite breathless from excitement and the exertion. He said: 'Did you see that? I thought for a moment it was all up with me. I tipped so, then so, and I threw out my legs thus and righted it. I have learned something new; I learn something new each time.'

Despite the obvious hazards, Wood was persuaded to have a go himself. He found it thrilling and decided there and then to buy one of Otto's gliders.

Otto's fourteen-year-old son had been with him when Wood was there. The following Sunday, 9 August 1896 there were no family members to witness his flights from Gollenberg Hill where Wood had photographed him. He travelled by train and wagon to the village inn on his own and there met his mechanic Paul Beylich. Just before he left he had a fierce argument with Gustav, who told him he should think of his wife and children and stop risking his life. But Otto would not abandon his passion. There were just a couple of locals watching as he made his first flight around midday. He made two further flights and then asked Beylich to take the stopwatch: with the breeze getting up, he thought he might get in a longer flight. It might have been in his mind, too, to try again the manoeuvre which had excited him when Wood was watching.

Lilienthal took to the air. All seemed well until the glider suddenly appeared to stop at a height of about 15 metres and then, as Lilienthal swung his legs in a desperate attempt to propel it forward, it went into a nosedive, propelling him head down and crashing to the ground and breaking up. When Beylich rushed to his aid with the other bystanders he found Lilienthal alive but apparently paralysed from the waist down. There was no medical help to be had nearby and a horse-drawn carriage took him to a hotel in the town of Stolin nearby, where he was seen by a doctor. Later that day Gustav was with Otto when he was taken to Berlin on a goods train, losing consciousness on the journey. In the morning he was taken in a horse-drawn ambulance to the clinic of a leading neurosurgeon but there was nothing that could be done and he died that afternoon, 10 August 1896 at around 5.30 p.m. The cause of death was given as fracture of the spine, but subsequent medical opinion suggests that he suffered a brain haemorrhage.

Paul Beylich had been working for Otto making his gliders. When he got to the factory after the inquest into the cause of the crash he found that everything they had been working on had been smashed up and burned in the boiler. Gustav, in his fury at what he regarded as the thoughtlessness of his brother, had ordered this, perhaps with the agreement of Otto's distraught wife Agnes. For a long time afterwards Otto's family would have a grim view of the 'dream of flight' and the selfish Icarus who had left them bereft and facing destitution. By contrast, in the world of aviation pioneers, Lilienthal attracted more attention in death than he had done when actively demonstrating his abilities with a glider.

* * *

Later in life, Wilbur said that he and his younger brother Orville had always done everything together and were inseparable. Both were very careful not to claim that one was more inventive or influential than the other. As their father had written after the success at Kitty Hawk, 'about equal credit is due each'. However, there is no doubt that it was Wilbur who first got interested in flying as a new kind of sport and that his inspiration was Otto Lilienthal. The 'Flying Man' was not that well known, but he had been featured in *McClure's Magazine* with dramatic mid-air photographs of his glides. It was a sport that required a good deal of courage, as well as dexterity and strength, the manoeuvring in flight achieved by dramatic shifts of the body. Gliding would be the ideal sport for the loner, and, by an unhappy accident, that is what Wilbur Wright had become as a young man in Dayton, Ohio.

In the winter of 1885–86, when he was nineteen, Wilbur had joined boys on a frozen pond in the gardens of the Soldiers' Home, a popular playground in Dayton, Ohio for a game of shinny, a rough and tumble kind of ice hockey. He was a good sportsman and gymnast, and academically ambitious: at Dayton school he was taking extra classes in trigonometry and Greek, hoping to get a place to study divinity at Yale. In an instant this bright future came to a shocking end. A hockey stick slipped from one of the boys' hands and smashed into Wilbur's face, knocking out several of his teeth. It took him a long time to recover from the accident. The disfigurement depressed him and he became afflicted with what his father, Milton, called 'nervous palpitations of the heart'. He abandoned school and sport and stayed at home where his mother, Susan, was terminally ill with tuberculosis. He nursed her until her death in July 1889 at the age of fifty-eight. When he had time he read through his father's extensive library, which included the *Encyclopaedia Britannica* and *Chambers's Encyclopaedia* as well as classics such as *The Decline and Fall of the Roman Empire*. For a while Wilbur helped his father, who was involved in a bitter internal battle in the United Brethren Church which led to a split and a breakaway group in which he became a bishop.

The first serious business enterprise of the Wright brothers was a printing works established by Orville and a friend. They attempted to publish newspapers, bringing in Wilbur as an editor, but the competition was too strong and they settled for the less glamorous work of jobbing printing: circus posters and the like. Orville was still a schoolboy but he had worked as a printing apprentice during the holidays and learned something of the business. In time they built their own press from odds and ends, which included an abandoned gravestone. In 1892 Orville bought a brand new bicycle for the huge sum of $160 and joined the craze that was sweeping America and Europe. Wilbur bought his bike a few weeks later at auction for $80. Shortly afterwards they decided to leave the printing works to be run by Orville's partner and to set up in the bicycle business.

Today it is not easy to imagine that the ability to ride a bicycle might have anything to do with the ability to fly, but in the 1890s it was commonplace to make the connection. James Means, successful manufacturer of the $3 shoe who became an enthusiast for aviation and published the short-lived, but hugely influential, *Aeronautical Annual,* wrote in the 1896 volume: 'To learn to wheel one must learn to balance. To learn to fly one must learn to balance.' Means invited Lilienthal to demonstrate his gliding in America but could not persuade him to leave his factory for any length of time. He published Otto's articles, however, and they wrote to each other. Otto agreed about cycling: 'I think that your consideration on the development between the flying machine and the bicycle ... is excellent. I am sure the flying apparatus will have a similar development.'

An article in the *Glasgow Herald* lamenting the death of Lilienthal in 1896 also made the point about the similarities of cycling and gliding:

> Falls must be expected in the preliminary trials. ... Similar difficulties have to be contended with when learning to ride a bicycle. The beginner is at first unable to keep his equilibrium, and so wobbles here and there, with the loss of much power, until he eventually finds himself hugging the earth. ... An adept rider ... quite unconsciously keeps his equilibrium without any exertion or loss of power. ... So it is with the new sailing machine, and it is only by practice that success can be attained.

The bicycles Wilbur and Orville bought in 1892 were the product of more than a century of innovation from the first hobby horses and boneshakers to the high-wheelers and finally the safety bicycles with a chain drive and inflatable tyres. Orville's *Columbia* was American made and featured at the World's Fair in Chicago which commemorated the arrival of Columbus in the New World in 1492. The advertising slogan of Pope Manufacturing Company of Boston was: *You have something to live for if you have not seen our new Century Columbia.* The bicycle industry was excitingly new and innovative, laying the foundation for many of the techniques used to promote the motor car which was following hard behind.

The brothers joined the Dayton YMCA Cycling Club and took part in July 1892 in the twelfth annual meet of the Ohio Division of the League of American Wheelmen which was staged in their town. It was probably the huge success of this event that persuaded them to go into the business of repairing and selling bicycles. In time they put together their own models, the Van Cleve and the cheaper St Clair, assembled mostly from parts manufactured elsewhere.

The bicycle business gave them a reasonable income as it grew, and they moved to larger premises in Dayton. But it was a while before Wilbur began to take seriously his long-held desire, or 'affliction' as he sometimes called it, to experience gliding. In retrospect the brothers always said that their interest had been quickened by news of Lilienthal's death. Wilbur read what he could in the home library. This included *Animal Mechanism* by the French Professor Etienne-Jules Marey who had become fascinated by the way birds flew and had developed a high-speed camera to capture the action. (See Chapter 2, on television.) However, it was not until nearly three years after Lilienthal crashed that Wilbur decided to write to the Smithsonian Institution in Washington for a reading list on the subject. His letter, dated 30 May 1899, was written on Wright Cycle Company headed notepaper and addressed to 'Dear Sirs':

> I have been interested in the problem of mechanical and human flight ever since as a boy I constructed a number of bats of various

> sizes after the style of Cayley and Penaud's machines (nb these are little ornithopters). My observations since have only convinced me more firmly that human flight is possible and practicable. It is only a question of knowledge and skill as in all acrobatic feats. Birds are the most perfectly trained gymnasts in the world and are specially well fitted for their work. . . . I believe that simple flight [gliding] at least is possible to man and that the experiments and investigations of a large number of independent workers will result in the accumulation of information and knowledge and skill which will finally lead to accomplished flight. . .

His letter asks for any references or documents the Smithsonian might have, while reassuring the Institution about the seriousness of his ambition: 'I am an enthusiast, but not a crank in the sense that I have some pet theories as to the proper construction of a flying machine.'

The reply came on 22 June from Richard Rathbun, Assistant Secretary of the Smithsonian with a list of works which included a quite extraordinary book entitled *Progress in Flying Machines*, published in 1894. Taking the story back to the machine designed by Leonardo da Vinci, the book listed every known would-be flying man and showed, with illustrations, how they had attempted to get airborne. Sir George Cayley was given some prominence. An article by Lilienthal was appended. The author of this work had the wonderfully evocative name of Octave Chanute. He was born in Paris in 1832, and his French parents divorced when he was six years old. His father, Joseph Chanut (Octave added an 'e' supposedly to Americanise his name), an historian and teacher, was appointed vice-president of Jefferson College and he took Octave with him to New Orleans. While he was still at school they moved to New York, where he had a private education. At the age of seventeen he found work as a chainman assisting with a survey for the Hudson River Railroad and began his self-education as an engineer. He was, amongst other things, a celebrated builder of railway bridges and his experience with those structures and the observation of the action of the wind on them appears to have triggered his interest in aviation. However, he kept his belief in the ability of man to fly mostly to himself during his hugely successful career, as he feared he would be regarded as a crank.

One of Chanute's last great engineering projects was the building of the stockyards in Chicago and when he retired there in 1890 he was at last able to indulge his interest in flying. A friend who edited the *Railroad and Engineering Journal* offered to publish Chanute's research into the history of aviation and the instalments appeared between November 1891 and December 1893. They were brought out as a book, *Progress in Flying*, three years later. At fifty-eight Chanute was too old to attempt to fly himself but he put together and funded a team of young men who would test gliders that he had designed himself. He found a test site in the sand dunes on the southern shores of Lake Michigan in Indiana and had some success, concluding that a biplane provided more lift than a single-winged machine.

When Wilbur and Orville received the Smithsonian reading list in Chanute's book they found meticulous research that provided them with just about everything that was known about every aviator in Europe or America. There was one, however, of whom they had heard but knew little about. He was an Englishman who was reportedly about to attempt to fly a powered machine. Percy Pilcher had been working as a naval engineer and lecturer when he read about Lilienthal's flights and thought he would like to have a go at building a glider of his own design. Born in England, his Scottish mother had taken the family to Germany when her second husband, Pilcher's father, died so he had learned something of the language. He was sent to boarding school in England when his mother died in 1877 making him an orphan at the age of ten. At thirteen he enlisted in the Royal Navy as a cadet, leaving when he was aged twenty to work as an engineer.

In 1895, when Pilcher was twenty-eight, he arranged to visit Lilienthal and to experience the *Fliegerberg* at first hand. In the weeks before he left he built his first glider, which became known as The Bat, a name suggested by the shape of the wings and the lack of a tailplane. He met Lilienthal in April and learned a great deal from him about his gliding technique; his grasp of German was a great help.

Pilcher built a series of gliders, and, lacking a convenient launch pad, developed the technique of towing his machines against the wind with a rope hauled by an assistant, and sometimes by a horse. He was the first to attach a wheeled undercarriage to a glider and he began to experiment with low-powered gas engines that might drive a propeller of sorts. Much of his experimental work was done in the grounds of Stanford Hall, the seat of Lord Braye whose son, the Honourable Adrian Verney-Cave, he had befriended in the Royal Navy. Pilcher was by all accounts a charming and retiring young man who was able to get funding from a number of influential people, one of whom was the Conservative MP for Canterbury, Henniker Heaton.

On 30 September 1899 Pilcher arranged for a demonstration of two of his gliders, one called *Hawk* and the other a triple-winged plane to which he attached a small motor. It was principally to show Heaton what he had achieved and the MP joined a distinguished group of guests including Lord and Lady Braye and Baden Baden-Powell, a man-lifting kite enthusiast and brother of Robert, founder of the Boy Scouts. His friend Adrian Verney-Cave was there, and Percy's sister Ella.

The day did not start well. There was heavy rain and the motor broke down before it could be tested in the machine. But Pilcher had the glider *Hawk* to show and arranged for it to be towed with a cotton line about 300–400 yards long. There was a gearing mechanism for multiplying the speed of drag, as the pace of a horse would not get him airborne. However, the first two attempts ended miserably when the line broke before Pilcher had gained more than 30 feet. At the third attempt he was rising as he had hoped when a sharp crack was heard as the tailplane broke away and Pilcher plunged to earth with a sickening crash. He was dragged unconscious from the wreckage and taken to a room in the Hall. He died there of his injuries thirty-four hours later.

The accident occurred while Wilbur and Orville were working their way through the reading list sent by the Smithsonian. It clearly gave them pause for thought. In May Wilbur addressed a letter to Octave Chanute in which he gave an account of his own ideas about aviation and asked for advice on a suitable place to practise flying. Written on Wright Cycle Company headed paper it begins:

> For some years I have been afflicted with the belief that flight is possible to man. My disease has increased in severity and I feel that it will soon cost me an increased amount of money and have been trying to arrange my affairs in such a way that I can devote my entire time for a few months to experiment in this field. My general ideas of the subject are similar to those held by most practical experimenters, to wit: that what is chiefly needed is skill rather than machinery. . .

Wilbur was not yet thinking of powered flight but of correcting mistakes he believed Lilienthal had made. 'I conceive that his failure was due chiefly to the inadequacy of his method, and of his apparatus. As to his method, the fact that in five years' time he spent only about five hours, altogether, in actual flight is sufficient to show that his method was inadequate.' He thought Lilienthal was wrong to try to regain equilibrium in the air by shifting his centre of gravity. 'My observation of the flight of buzzards leads me to believe that they regain their lateral balance, when partly overturned by a gust of wind, by a torsion of the tips of the wings. If [the] rear edge of the right wing is twisted upward and the left downward the bird becomes an animated windmill and instantly begins to turn, a line from its head to its tail being the axis. . . . In the apparatus I intend to employ I make use of the torsion principle. . .'

In order to experiment with this principle, Wilbur said, he would build a tower from which he could be suspended by a kind of safety rope while he tested the effect of the wind on his machine. That way he could stay in the air for hours and would gain much more experience than if he tried free flight. 'The method employed by Mr Pilcher in towing with horses in many respects is better than I propose to employ, but offers no guarantee that the experimenter will escape accident long enough to acquire skill sufficient to avoid accident.' At the end of the letter he asked if Chanute had any more information about Pilcher.

Wilbur had set his mind to learn from the mistakes of others and, with Orville, had begun to formulate an approach to controlled gliding which would be critical to their later success. Lilienthal's method of shifting his body to balance his glider was clearly unsatisfactory and evidently extremely dangerous. The Wright brothers would devise a mechanism whereby their wings could be twisted to compensate for uneven wind pressure. Whereas Lilienthal and Pilcher had hung from the glider frame, their legs dangling, their bodies used for 'steering', the Wright approach would be to lie flat across the wing, which aerodynamically was more satisfactory. The shape of the wings, the curvature, the relationship between the length and the width, would be crucial too. They had Lilienthal's calculations relating lift and drag and they would test these.

It was not long before Wilbur had a reply from Chanute. He would be only too pleased to help with information about Pilcher, though this was scant. They were directed to more articles and, in particular, to the kites designed by an Australian, Lawrence Hargrave. Born in England in 1850, Hargrave joined his father and older brother in New South Wales when he was fifteen years old. After an adventurous young life in which he was an explorer and marine engineer he became independently wealthy on the death of his father. Always fascinated by the force of wind and waves, he became interested in aviation and began to experiment with lightweight steam engines. His study of man-lifting kites led to his invention of the box kite, which was adopted by the US Weather Bureau to carry measuring instruments to high altitudes. Hargrave managed to raise himself 16 feet off the ground with a string of box kites flown in a 20 mph wind.

Chanute was a great champion of Hargrave and it is sometimes suggested that his box kite influenced the Wright brothers when they were designing their flyer. The brothers denied this, not because they intended to denigrate Hargrave or to avoid a patent dispute: Hargrave refused to patent any of his inventions on principle as he thought it was a form of robbery. They did have some correspondence with him, and his box kite principle was certainly incorporated into some of the earliest

European powered aircraft. But the Wright brothers never adopted untried or untested any component borrowed from an existing machine: this was one of the secrets of their ultimate success for they discovered flaws in the calculations of their predecessors, which they then corrected following their own research. For them Lilienthal was the giant of aviators because he tested his theories of aerodynamics by sailing into the wind. And that is what they were about to do. There was a sense of urgency as they were aware that elsewhere in America there were rivals much more distinguished than themselves intent on being the first to fly a piloted, heavier-than-air machine.

* * *

The outbreak of war with Spain in April 1898 persuaded the United States Government to take seriously a claim by Samuel Pierpoint Langley, a distinguished astrophysicist and Secretary of the Smithsonian Institution, that he was close to being able to offer the military a revolutionary piece of equipment: a motor-powered aeroplane, or, as Langley liked to call his flying machines, an aerodrome. On the face of it Langley's proposal appeared reasonable. Twice in 1896 he had been able to fly a model 'aerodrome' powered by a miniature steam engine. These test flights had been from the roof of a houseboat moored on the Potomac River, the model planes launched with a catapult mechanism. If the claims of the Englishman Stringfellow to have flown a powered model earlier are discounted, then Langley had a first in aviation history. All he needed to build a full-sized, man-carrying version of his aerodrome was sufficient funds to develop a more powerful engine that did not weigh too much.

The War Department's Board of Ordnance and Fortifications investigated and, despite the fact that there was still much scepticism about heavier-than-air flight, reported favourably. As it turned out, the war with Spain lasted only ten weeks but in November that year Langley still got a grant of $50,000 to be paid in two equal instalments. The *Washington Post* got hold of the story and commented impudently:

'We will do our best to be at the launching. We intend to practice longevity with the most industrious enthusiasm in the meanwhile. We should never forgive ourselves were we so careless as to die before the ceremony.'

To look after the project Langley took on Charles Manly, a graduate majoring in electrical and mechanical engineering at Cornell University, who had been recommended as a young fellow with 'gumption' and some technical training. Searching for someone to build the kind of engine he needed he found Stephen Blazer, a Hungarian immigrant watchmaker who had built the first motor car in New York. Langley wanted a rotary engine of a certain weight and power, and Blazer blithely said he could build one. It turned out to be a tougher assignment than he had imagined and the first design was a failure, generating far too little power.

With his mind set firmly on the engine for his giant aerodrome, Langley set off for Europe with Manly to consult some of the more prominent engineers. In England they called on the American Hiram Maxim who had made a fortune from the sale of his machine gun and was knighted for his efforts. Maxim spent around £30,000 building a huge machine with propellers powered by steam engines. It was not really designed to fly; it was more of an experimental craft to see if such a heavy machine with men on board could get off the ground. It was designed to run on a track and did briefly take to the air before crashing. Maxim told Langley that Blazer's engine would be no good. So did other authorities. Manly returned to Blazer's engine and began to modify it. Blazer himself, still trying to fulfil his commission from Langley, went bankrupt.

It was while Langley was desperately trying to find an engine for his giant aerodrome that the Wright brothers decided that they would spend some time away from their bicycle business in order to experiment with gliders. Whereas Langley imagined that the discovery of a powerful, lightweight engine was all that was needed to take to the air, the Wright brothers went back to the drawing board and the model of the bird that had inspired Otto Lilienthal.

* * *

Meticulous as always in his research, Wilbur had written to a few weather stations on both the east and west coasts to enquire about the local terrain and weather conditions. The brothers had chosen Kitty Hawk in North Carolina not only because this bleak coastal area seemed to have all the elements they needed but also because of the encouragement of the local people. Wilbur had received positive and helpful letters both from the head of the weather station and from Bill Tate who, with his wife Addie, ran the local post office.

Half the community in Kitty Hawk were still illiterate but Bill had had some schooling and had a nice, evocative turn of phrase: 'If you decide to try your machine here & come, I will take pleasure in doing all I can for your convenience & success & pleasure, & I assure you of a hospitable people when you come among us.' As well as promising Wilbur a warm reception he provided an enticing description of the local terrain:

> At Kitty Hawk there is a strip of bald sand beach, free from trees, with practically nothing growing on it except an occasional bunch of buffalo grass. This strip of beach is about 1500 yards wide from ocean to bay, and extends many miles down the coast. The average elevation is from 8 to 20 feet above sea level, but at certain places sand hills have been piled up by the wind until some of them (the Kill Devil Hill) have reached an elevation of 75 to 100 feet above the plain. . . . The prevailing winds are from the northeast, and these hills are very steep on the southwest side, but not so steep on the northwest side. They average from 20 to 45 degrees on the south side.

That description made up Wilbur's mind. It sounded ideal. However, he had not asked how to get there and Bill Tate and the people at the weather station did not know if he was going to turn up. It turned out that the last leg of the journey from Dayton would have to be by boat, and the regular ferry ran infrequently. Wilbur persuaded a fisherman to take him in a boat that leaked alarmingly and was laid up on the way when the weather turned rough. On 13 September 1900 Wilbur arrived at the Tates' roughly furnished timber house having eaten nothing for two days. Addie Tate lit a fire and cooked him eggs and bacon. It was the first of many meals they shared and the start of a collaboration between the Wright brothers and the fishermen and weathermen and lifesavers in the quest to get a powered aeroplane to fly. Without the people of Kitty Hawk the brothers would have been unable to achieve much more than a bit of speculative kite flying. Bill became a lifelong friend.

When he had had his breakfast with the Tates Wilbur asked where he could find lodgings. The Tates went into a huddle to discuss what they might have to offer and Wilbur overhead their discussion about whether or not to take him in. He was clearly a much more sophisticated person than they were used to meeting. Wilbur interrupted to reassure them that he would make no demands and offered to pay his way. Bill Tate recalled later that the only stipulation made was that he should boil some water from their primitive well for their guest as he had a terrible fear of typhoid, from which his brother had suffered. Orville would be arriving in Kitty Hawk in two weeks' time bringing with him a tent and more material for the glider. Wilbur had enough to be getting on with and began to assemble their first glider in the front yard of the Tates' house. Addie Tate's sewing machine came in handy when some of the fine French sateen used for the wings had to be re-sewn.

Ten days after they got to Kitty Hawk, Wilbur wrote to his father to reassure him he had arrived safely and to outline his plans. By now he was contemplating the time when he might try to build a powered machine. But that would have to wait.

> I have my machine nearly finished. It is not to have a motor and is not expected to fly in any true sense of the word. My idea is merely to experiment and practice with a view to solving the problem of equilibrium. I have plans which I hope to find much in advance of the methods tried by previous experimenters. When once a

> machine is under proper control under all conditions, the motor problem will be quickly solved. A failure of the motor will then mean simply a slow descent & safe landing instead of a disastrous fall.

This insight was to be the key to the Wright brothers' success. They were, to use the term coined by the late Charles Gibbs-Smith, a distinguished British aviation historian, true 'airmen' in contrast to those inventors, such as Samuel Langley or Hiram Maxim, he called 'chauffeurs'. In *The Aeroplane, an Historical Survey* he wrote:

> The chauffeur attitude to aviation regards the flying machine as a winged automobile, to be driven into the air by brute force of engine and propeller, so to say, and sedately steered about the sky as if it were a land – or even marine – vehicle which had simply been transferred from a layer of earth to a layer of air. . . . Whereas the airman thought primarily in terms of control in the air, and quickly realised the unpowered glider was the vehicle of choice.

On that first trip to Kitty Hawk the brothers lived in a tent after they had moved out of the Tates' house, and they treated the trip almost as a break from the bicycle business: Wilbur in one letter referred to it as a holiday. They soon abandoned the scheme of flying their glider-kite from a tower and simply flew their glider as a kite most of the time, testing what they regarded as their most important innovation: the twisting of the tips of the machine in imitation of the buzzard's wings when it righted itself. This became known as wing warping, for which they would require a specially constructed set of levers. Most important, though, was the shape and size of the wings. There was not a lot to go on apart from the calculations made by the Lilienthals. At first they relied on these, but in the first two trips to Kitty Hawk in 1900 and 1901 they began to realise that the equations which predicted the amount of lift they would get with a given area of sailcloth and wing shape were giving disappointing results.

Back in Dayton they would try to figure out what was wrong with their designs. All the while they discussed their discoveries with Octave Chanute. He encouraged them when they became despondent, arranged for Wilbur to give a talk to the Western Society of Engineers, and sent out fellow aviators to Kitty Hawk to provide some assistance. At one point Chanute offered to fund the brothers, but Wilbur refused to hear of it.

In their trips between Kitty Hawk and their bicycle business in Dayton between 1900 and the end of their stay in 1902, Wilbur and Orville in effect relived in a short period every stage in the evolution of manned flight from the early efforts of Sir George Cayley up to the tragic deaths of Lilienthal and Pilcher. When they were convinced that Lilienthal's tables of lift and drift were not accurate enough they had begun to conduct tests themselves. One ingenious device was a wheel mounted horizontally on the front of a bicycle so that it would be spun by the flow of air as the rider got up speed. Attached to the wheel were two blades, one wing shaped and the other flat, and by adjusting them they began to get a crude idea of the effect of the play of air on different settings. When the bicycle tests suggested that the Lilienthal calculations were wrong the brothers devised a wind tunnel, a simple one in a box at first and then a larger device in which the airflow was generated by a fan spun by a gasoline engine.

In 1901 gliding had presented them with a number of puzzles. They could not explain the sudden tilts and dives of the gliders. However, by meticulous re-examination of the forces at work and with a more sophisticated idea of the most favourable size and shape of the wings they returned in 1902 to the sand dunes of Kitty Hawk feeling much more confident. They built a more permanent camp with the help of a local carpenter and had a number of Kitty Hawk helpers on their payroll. Bill Tate had also insisted they pay a nominal rent to the family who owned this desolate piece of coastline; otherwise they could be evicted and their camp torn down. For the most part they got on very well with the local people, most of whom, according to Tate, thought they were plumb crazy. The brothers were always polite and considerate and often cooked meals for those who helped them launch the glider and got in their supplies. They brought the first gasoline stove to Kitty Hawk and made a gift of it to Tate. And on their return in 1901 they had found the Tates' two little daughters in dresses made from the sateen wing material of the first glider.

* * *

The Wrights had brought a new economy with them to Kitty Hawk and were rewarded with an enthusiastic team of helpers whose boats and horses kept them supplied with food and materials. The trek from the little settlement to their camp over the dunes was hard going. The wildlife was troublesome too. Not just the thick swarms of mosquitoes but wild hogs, which would raid the camp, and mice that nibbled their faces at night. There was some fun when they took part in shooting contests and hunted game. As the birthplace of a remarkable modern invention it could not have been more bizarre.

In most photographs of the glider launches there is a familiar figure who spent most of his time at the camp that year doing all kinds of jobs when he was not needed to hang on to the glider. This was Dan Tate, Bill's half-brother, a fisherman during the season on the coast. He earned a regular salary from the Wrights but would return to fishing at the end of October. When Dan left, the Wrights packed up and returned to Dayton. By the end of their stay at Kitty Hawk in 1902 they became confident – after a last session of 250 glides in all kinds of wind conditions – that they were able to control the glider with sufficient confidence to attempt powered flight. Back in Dayton they were intent on designing two entirely new pieces of equipment: a lightweight motor and propellers. There were plenty of gasoline engines about, used both to drive machinery and to power the first motor cars. They sent their specification to a number of manufacturers but none would attempt to build an engine that gave eight horsepower and weighed just 180 pounds. So they had to design one themselves. Nobody was making propellers except for boats.

It was fortunate that the brothers had hired in June 1901 a self-taught engineer, Charles E. Taylor, to handle their repairs when they were away at Kitty Hawk. Charlie (as he was known) took up the challenge of building an internal combustion engine from scratch enthusiastically, working on the design with the brothers and making many of the parts himself. The town of Dayton was full of small manufacturing firms and the Buckeye Iron and Brass Foundry was able to make their crankcase out of lightweight aluminium. Starting in the winter of 1902–3 Charlie had the engine ready to run within six weeks. It was more powerful than anticipated, allowing the brothers to increase the weight of their flyer.

When they came to design the propellers, the Wrights could find no useful theory or principle to help them. From a study of the evolution of a ship's propellers they found that there were no sophisticated designs; it had mostly been a matter of trial and error. So, too, with the massive propellers that were supposed to lift the machines designed by Langley and Maxim. The Wrights carried out tests in their wind tunnel and came to the conclusion that they could apply the same principles to the blades of the propeller as to the design of the wings. They would have two propellers rotating in opposite directions.

The experience Wilbur, Orville and Charlie had in repairing and building bicycles was invaluable. To harness the motor to the propellers they could cannibalise bicycle gears and chains. To balance and turn in the air they needed to bend the tips of the flyer's wings: bicycle brake wires, chains and cogwheels could be adapted for flight control. They did not attempt to create landing gear with bicycle wheels but they used their own hubs to hold the machine on its single launch rail. The wing structure of the flyer incorporated some American wooden bridge-building technology. And then there were the photographs.

When they began their gliding at Kitty Hawk the Wrights bought one of the most expensive cameras available, a huge contraption which stood on a tripod. It was a Korona-V and cost $85. They took many pictures of the people of Kitty Hawk, but for them photography was more important as a means of recording their experiments. They could not develop the plates at Kitty Hawk and had to wait to see what they had captured when they got home to Dayton. Here in their darkroom they had the thrill of reliving their glides out on the sand dunes. In a lecture Wilbur gave in 1901 to the Western Society of Engineers he showed lantern slides made from their photographic record of glides to illustrate this and explained how difficult it was then to get a moving image on film. Turning to one picture of a glide he remarked: 'In looking at this picture you will readily understand that the excitement of gliding experiments does not entirely cease with the breaking up of camp. In the photographic darkroom at home we pass moments of as thrilling interest as any in the field, when the image begins to appear on the plate and it is yet an open question whether we have a picture of a flying machine, or merely a patch of open sky.' The brothers would take their Korona-V with them when they returned to the North Carolina coast on 25 September 1903, ready to assemble the powered flyer.

For the Wright brothers photography was another technological innovation they turned to their advantage. For the failed aviator, however, the camera was not so much a boon as a newly invented means of humiliation.

* * *

On 7 October 1903 Charles Manly attempted to fly Langley's 'aerodrome', launching it from the same houseboat on the Potomac that had been the scene of the successful model flights. It was to be a huge event, a world first, and the newspapers were gathered to record it. They had had to wait while repairs were made and were becoming impatient. Manly started the engines and when his power was up a catapult mechanism launched him skyward. The photographers captured an image of the aerodrome as it nose-dived directly into the Potomac with Manly still aboard. Charles Gibbs Smith in *Aviation* commented:

> It was the brave Manly who volunteered as that pilot. The extent of such monumental folly may be truly gauged if one remembers that Manly had never even tried to fly a glider, let alone a completely untried powered machine, and he was now to be precipitated into the air at some 30mph without the slightest idea of what would happen when he got going: he would have a rudder and an elevator, but no experience in the cockpit of working them, and no idea at all as to how they would affect the flight of the Aerodrome . . . it seems incredible to us today that Langley should have been prepared to risk a man's life in this way.

Wilbur Wright read the news of the abject failure of Langley's machine in a newspaper cutting Chanute attached to a short letter he sent from Chicago to Kitty Hawk on 7 October. It must have been an evening paper that had the headline: 'Prof. Langley accused of using government time and money on visionary flying machine: congressional investigation likely.' An account of the embarrassing failure followed. Wilbur, replying to Chanute from Kitty Hawk on 24 October wrote: 'I see that Langley has had his fling, and failed. It seems to be our turn to throw now, and I wonder what our luck will be. We will still hope to see you before we break camp.'

* * *

It was not plain sailing for the Wrights either. When they had returned to Kitty Hawk on 25 September they immediately began to make repairs to their hut and workshop of the previous year and to enlarge it and make it more comfortable. Dan Tate and others had already brought timber in as well as groceries. The wild hogs had gone and the mice appeared to have died of starvation while they were away. But the shifting sands had moved their camp around a bit and they soon discovered how the landscape they had chosen was sculpted by the weather when a tremendous storm very nearly blew them away. Wilbur sent home a little sketch of Orville trying to hammer nails into the roof of their hut with his overcoat tails blown over his head. It was a terrible year for storms and they had to delay their planned flights with the powered machine. They fell out with Dan Tate, who refused to do something they had asked and complained that his wage of $7 a week was too little. He took his hat and left the camp.

However, it was the treacherous weather and shifting sandbars on this North Carolina coast which had persuaded the government to establish several life-saving stations, including the one at Kill Devil Hills where the Wrights were flying. And this turned out to be crucial for their eventual success. The station got a new keeper-in-charge in 1899, Jesse Etheridge Ward, a native of Roanoke Island in his early forties who had been a fisherman before he joined the Life-Saving Service. One of Ward's crew, Adam Etheridge, visited the Wrights as they were setting up and repairing their camp and began a friendship with the surfmen. The crew of the Kill Devils Hill Life-Saving station brought the brothers their mail and would stop to watch them work on the glider. 'We assisted in every way and I hauled lumber for the camp,' Adam Etheridge recalled later. 'In pretty weather we would be out there while they were gliding watching them . . . we began to become interested in carrying the mail just to look on and see what they were doing.' The Wrights' camp could be seen from the life-saving station and they came to an arrangement with Captain Ward (a title his men awarded him) that if crewmen were needed to assist them they would raise a red flag. Any men who could be released were given permission to head over to help set the flyer on its course.

Delayed by the weather, they did not test the engine harnessed to the propellers until November. On the first run the engine backfired and the jolt broke one of the crankshafts. They had to send back to Dayton to get stronger crankshafts made, and it was two weeks before they could try again. As soon as they ran the engine with the propellers another problem arose. The sprockets on the shafts kept coming loose. Orville recalled how they solved the problem: 'While in the bicycle business we had become acquainted with hard tile cement for fastening tires on the rims. . . . We heated the shafts and sprockets, melted cement into the threads, and screwed them together again. This trouble was over.'

But there were more problems. Another shaft in the machinery cracked and Orville had to go back to Dayton for replacements. Again this took two weeks and the machine was not ready for another test until 12 December. That was a Saturday, and though the brothers were intent on trying the flyer as soon as possible they refused to work on Sunday and spent most of the day in their hut, reading. The next day the weather was not ideal but they decided to attempt a flight. A mile away, Captain Ward's crew spotted the red flag flying to indicate that the brothers needed help. As always, Wilbur and Orville were dressed in starched collars, ties, suits and hats when they greeted the surfmen John T. Daniels, Robert Westcott, Thomas Beacham, W.S. Dough and 'Uncle Benny' O'Neal who answered their call. Without them it would have been quite impossible to get the flyer in to position on Kill Devil Hill. The machine weighed 750 pounds and could not move under its own power. The men laid out the 60-foot launch rail and slid the machine along it, then moved the rail to the front and repeated the move all the way to the hill. It took about forty minutes.

'We took to the hill,' Wilbur wrote home that night, 'and after tossing for first whack, which I won, got ready for the start.' The wind was really too light, the track not laid quite right, and, as Wilbur admitted, he made an error of judgement at the take-off. But they were far from despondent. 'There is now no question of our final success. The strength of the machine is all right, the trouble in the front rudder being easily remedied. We anticipate no further trouble in landings.' The flyer was soon repaired and they were ready to raise the red flag again on 17 December. Daniels and Dough were there, as was Adam Etheridge, a man called Brinkley from the town of Manteo and a boy, Johnny Moore.

It was Orville's turn to fly, but before he mounted the machine he set up his Korona camera and positioned it to capture the moment of take-off. Wilbur would be alongside the machine so he asked Daniels to squeeze the bulb that would work the shutter. The Kill Devil Hill surfman got it right, snapping one of the most famous photographs in the world: Orville and Wilbur's 'eureka moment'. The fact that it was such a short flight – just 12 seconds – did not disconcert the brothers. They had already beaten Langley. The fourth flight, of nearly a minute, was almost too good to be true. And they would be home for Christmas.

* * *

When they decided to take the risk with the flying machine they agreed that they would treat it strictly as a business: they would, sometime, want to get their money back. When they looked for a testing ground closer to home the best they could find was a cow pasture known as Huffman Prairie owned by a Dayton Bank president, Terence Huffman. They got it rent free on condition that they herded the cattle away from the flight path of the machine. In the spring of 1904 they built a workshop there and invited newspapers to witness their first efforts. These were failures and the press lost interest. However, little by little, they improved the performance of the machine and pursued their ambition to be able to fly in a complete circle. If they could do this they would be flying like a bird, proving to themselves that they had a real mastery of the air.

The French believed they had a kind of historical right to invent the first working aeroplane. In the eighteenth century the Montgolfier brothers had been the pioneer balloonists. On 19 October 1901, Santos Dumont had won the prize offered for the first to fly the twelve-mile round trip from the district of St-Cloud to the Eiffel Tower and back in less than thirty minutes. His airship was a dirigible; it could be steered, and was powered by a propeller. Santos Dumont was adopted by the French, though he was a Brazilian who had been sent to study science in Paris in 1891 at the age of eighteen. He endeared himself to his adoptive country when he donated the 100,000 francs he won to his staff and the beggars of Paris. In Europe there was far more interest in airships than in heavier-than-air planes. Santos Dumont learned about the Wright Brothers' flights from Octave Chanute when they met in

St Louis in 1904 and was one of the first in Europe to try to emulate their success.

The brothers had spent 1904 improving the performance of their flying machines at Huffman Prairie within easy reach of Dayton but the advances they made did not attract much attention either from local people or the newspapers. In fact nothing of their success was reported until the arrival at the Wrights' home in Dayton one day of a remarkable man driving an Oldsmobile Runabout who asked if he might witness one of the flights he had read about. This was Amos Ives Root, who had driven across Ohio from his home in Medina, partly for the novelty of motoring at a time when most vehicles were still horse drawn, and partly to explore the countryside. Root was himself a great innovator, taking an early interest not only in the natural world when he worked in the family's market garden, but in electricity and anything mechanical. He learned to make jewellery and built a factory to manufacture it. But his greater claim to fame grew from his hobby as a beekeeper. He invented a system for taking honey from hives without damaging them and sold the equipment around the world. He also published a magazine, *Gleanings in Bee Culture,* in which he wrote about all kinds of subjects. The first reports of the Wright brothers' Kitty Hawk flights had attracted his attention. Now he had come to see for himself what they were up to.

As they had improved their flying machine, Wilbur and Orville had become more secretive about its design and when Root introduced himself they were no longer inviting journalists to report on their efforts at Huffman Prairie. However, they clearly recognised in Root a kindred spirit and invited him to watch them fly. On 20 September 1904 Root was astonished to witness Wilbur's first successful circular flight. He made extensive notes of his impressions and asked if he could publish a piece in his magazine. The brothers asked him to hold back publication for a while so they could replicate that epic flight. Root's account eventually appeared in the 1 January 1905 edition of *Gleanings in Bee Culture* under the heading 'What Hath God Wrought?', the same biblical quote that was chosen for Samuel Morse's first official American telegraph message sent in 1844 to inaugurate the Washington to Baltimore line.

Root's writing style was eccentric as he struggled to capture the sheer wonder and excitement of Wilbur's flight at Huffman Prairie. How to describe it to people who still believed that what he had seen was impossible?

> When it first turned that circle, and came near the starting-point, I was right in front of it; and I said then, and I believe still, it was one of the grandest sights, if not the grandest sight, of my life. Imagine a locomotive that has left its track, and is climbing up in the air right toward you – a locomotive without any wheels we will say, but with white wings instead . . . imagine this white locomotive coming right towards you with a tremendous flap of its propellers, and you will have something like what I saw.

An evangelical Christian, Root believed invention was God-given and that the human race had a duty to discover new things. For him, watching Wilbur turn a full circle in the air was a deeply religious experience:

> God in his great mercy has permitted me to be, at least somewhat, instrumental in ushering in and introducing to the great wide world an invention that may outrank the electric cars, the automobiles, and all other methods of travel, and one which may fairly take a place beside the telephone and wireless telegraphy. Am I claiming a good deal? Well, I will tell my story, and you shall be the judge.

Root offered the readers of *Gleanings of Bee Culture* a reasonably sophisticated account of how the flying machine took to the air, the shape of the wings providing lift in the same way as the gliding bird moves with just the slightest adjustment of its wings. But when he offered the piece to *Scientific American* they turned it down. Was Root's prose too flowery and too evangelical in attributing the successful flights to God's providence? Or did the editors simply disbelieve his story? It seems that in 1905 the world was not ready to accept that powered, manned, heavier-than-air flight was possible. The fact that the Wright brothers had no academic qualifications and were allegedly mere bicycle mechanics and manufacturers did not help their credibility. Yet nobody in America – and no one else in the world – was more knowledgeable about the history of attempts to fly than Wilbur and Orville. As Root noted: 'When I first became acquainted with them, and expressed a wish to read up all there was on the subject, they showed me a library that astonished me; and I soon found they were thoroughly versed, not only in regard to our present knowledge, but everything that had been done in the past.'

When they were confident they had mastered powered, piloted flight, the brothers decided to mothball the aeroplane as they feared their invention might be too easily copied. Newspaper interest quickened but those who wanted to verify their achievements had to resort to tracking down witnesses. In 1906 the brothers finally got a patent for their glider, but not for the powered vehicle. They were then intent on recouping the money and time they had invested.

For two-and-a-half years they stopped flying altogether while they waited for a firm order that would give them the funds to continue the production of their machines. In 1907 they travelled to England and then to France, where Wilbur had his first flight in a balloon. European aviators who were still failing to take off in a great variety of flying machines were blissfully unaware of what the Wrights had achieved. The British were doing nothing, while the French, who were confident they were in the vanguard, had managed only a few short hops in their 'chauffeur' style machines.

The brothers failed to sell their invention to the military but a great deal of interest was shown by businessmen. In anticipation of French orders they crated a Flyer to France and had it stored. Orville ordered some French petrol engines. On the same trip the Wrights went to Germany but they returned to Dayton empty handed. Just when it looked as if the whole enterprise would fail they got not one, but two, orders. The US War Department invited bids from anyone who could make a flying machine that could for one hour carry a pilot and passenger weighing not less than 350lb, show an average speed of 40 mph in a ten-mile test and carry enough fuel for 125 miles. Though the proposal was ridiculed in the newspapers there were forty-one bids. The Wrights won with a tender of £5,000, with delivery promised in 200 days. At the same time they sold the French rights to a wealthy industrialist for 500,000 francs in cash and half of the founders' shares on condition that they could make two public demonstration flights of 50 kilometres in a time limit of one hour. The flights had to be made within four days of a pre-arranged date, which should be no later than five months after 1 June 1908.

Wilbur wrote to Chanute to outline the French deal saying they would have preferred to tackle the US War Department bid first but they really could not turn down the French offer, which might not be repeated. This sudden success presented them with a huge challenge. They had not flown for more than two years and felt they would be rusty. To satisfy the demands of both their clients they would have to make practical demonstrations at somewhere near the limits of their most recent achievements. They decided to return to Kitty Hawk to get in some flying practice away from the public eye. Wilbur arrived there on 9 April to find their camp derelict and looted; he was joined a week later by a mechanic Charles Furnas and by Orville on the 25th.

When they had rebuilt the camp and began to test-fly their latest machines, at first using sandbags for passengers, they began to notice furtive movements in the brush of Kitty Hawk's sand dunes. An episode unfolded which must be one of the most bizarre in the history of technological innovation: along with the wild hogs and buzzards and mosquitoes, Kitty Hawk sands had been invaded by a swarm of paparazzi. Just as the brothers were preparing to demonstrate their superiority in the air, there were reports from France that the first European aviators had managed to get off the ground and fly for a kilometre or more. This finally convinced the newspapers on both sides of the Atlantic that the aeroplane was not a mythical creature but a reality. The Wrights were tracked down to Kitty Hawk by journalists, who assumed the Wrights would refuse to fly if they showed themselves. From the insect-infested brush that bordered the dunes the newspaper spies watched in amazement as the Wrights made practice runs with a passenger on board.

They used sandbags at first to make up the weight, then Orville carried the mechanic, Charles Furnas: the brothers had promised their father they would not fly together, just in case. When Wilbur crashed the Flyer their experimentation had to come to an end. But he was not badly hurt and the brothers left Kitty Hawk confident that they were ready to make the flights that would fulfil the terms of their contracts in America and France.

Wilbur went straight from Kitty Hawk to France where they had left a crated flyer the previous year. Meanwhile Orville with two engineers prepared to put the flyer through its paces at the US Army Signal Corps base at Fort Myer, Virginia. Wilbur was the first to fly on a racecourse close to Le Mans where the motor car manufacturer Leon Bollée had given him factory space to assemble the machine. In July he had been badly scalded while testing the cooling system of the engine, and with the delay in flying French newspapers noted sceptically 'Le bluff continue'. They looked forward to yet another comical aviation display. It was a cynical gathering of journalists and leading French aviators that assembled on 8 August at the Hunaudières racecourse to watch Wilbur prepare for his first European flight. The flyer was launched from a rail by a rope attached to a large weight suspended from a gantry. As always, Wilbur wore a high starched collar, a grey suit and a cap. It was a very small area in which to fly and he was airborne for only 1 minute 45 seconds. But the impact was sensational. Nobody had seen anything like it before and the French had to eat their words – in particular the newly coined Anglicism *bluffeur*. It was the start of an astonishing European tour that would establish the Wrights as international heroes.

Orville's first flight was on 3 September. This too was a sensation. He broke his own records for flights lasting more than an hour and took up a number of passengers. The American demonstrations, however, came to a tragic end when Orville crashed and his passenger, Lt Thomas Selfridge, was killed, his skull cracked when his head hit one of the spars. Orville himself was badly injured and was unable to fly again that year, and got about on crutches. However, Wilbur needed his brother to supervise work in France and in January 1909 Orville and their sister

Katharine (nicknamed Sterchens) arrived in Paris and then travelled to the Pyrenean winter resort of Pau where Wilbur had moved. This became a Mecca for royalty and celebrities who all wanted to experience a flight in an aeroplane.

The success of the Wright brothers prompted Gustav Lilienthal to take a more sanguine view of flying than he had when he had Otto's gliders destroyed. He tried to form a company to investigate 'safe flying' and wrote to James Means, the Boston businessman who had tried to persuade Otto to go to America, to suggest that a Lilienthal Society might be formed. He took the opportunity to point out to Means that Otto's family were living in poverty and wondered if some of his rich friends might help them out. Otto's widow Agnes had to care for a son who was incapable of working.

When Wilbur Wright visited Berlin in April 1911 he took the opportunity to pay his respects to his great hero, Lilienthal. He wrote to Orville: 'Frau Lilienthal is a nice looking and very intelligent woman. I liked her very much. One of her sons is not quite right in his head. Her daughter is married. The other son is just out of school and seems to be a very nice young man. If we get a pile out of France I am going to give some of it to them. We rode down to their old home, and also down to the hill which Lilienthal built. It is really quite a suitable monument to his memory and will probably be there after most other monuments have perished.' In December that year Agnes received a letter from Dayton, Ohio:

> Dear Madam:
> As you already know we have great admiration for the work of your deceased husband the late Otto Lilienthal, and it has been a matter of much regret to us that we never had the pleasure of his personal acquaintance. He was a great man.
>
> As a token of our appreciation of him we beg that you will accept the enclosed exchange for one thousand dollars with our best wishes for a Merry Christmas and a Happy New Year.
> Yours truly
> Wright Bros

The following April Wilbur became sick while on a trip to Boston. He was exhausted and worn down with constant patent disputes. Back in Dayton the brothers had just acquired a large plot of land and had commissioned architects to build a house. After a picnic there in May Wilbur ran a high temperature. The doctors did not know what it was, but feared it was typhoid, which had nearly killed Orville in 1896. Wilbur did not recover and on 30 May 1912 he died at the age of forty-five.

Gustav, meanwhile, began to develop a flying machine with flapping wings. It was a bizarre enterprise and a complete failure. Otto's *Fliegerberg* was made a monument in 1932 in a resurgence of German national pride. The following year Gustav died of a heart attack. The Nazis had just come to power and they commandeered the funeral for their own propaganda purposes. Hitler Youth accompanied the coffin and a wreath from Hermann Goering was dropped from an aeroplane circling the cemetery. It missed the coffin by a long way, much to the relief of his widow Anna and her daughters. She turned down an invitation to join the Nazi Party.

Thousands had paid their respects to Wilbur at his funeral in Dayton. Nobody doubted that he and Orville were the true inventors of the aeroplane. But there was a campaign afoot to deny them that honour. Samuel Langley had died in 1906 a disappointed man after the failure of his aerodrome and the clear superiority of the Wright flyer in 1903. A rival of the Wright Company in America, Glen Curtiss, plotted to resurrect the 1903 aerodrome and to show that it could have flown if the launch mechanism had not been faulty. As Langley had 'had his fling', as Wilbur put it, before the Kitty Hawk flights, this could undermine the Wright brothers' patent. Curtiss and his conspirators at the Smithsonian Institution completely rebuilt and modified the aerodrome and in 1915 managed to fly it a short distance. The original 1903 aerodrome was then exhibited in the Arts and Industries Building of the Smithsonian with the caption: 'The first man carrying aeroplane in the history of the world capable of sustained free flight.' Orville was understandably upset. The feud dragged on for years, long after he had sold the Wright Company. It was not until 17 December 1948 that the 1903

Wright flyer took its rightful place in the Smithsonian, exactly fifty-five years after the first flight at Kitty Hawk. Orville had died in January of that year at the age of seventy-six but he had relented and changed his will, so the Flyer was bequeathed to the Smithsonian. From 1925 it had been exhibited in the Science Museum in London and it was still there when the Luftwaffe began their raids in 1940. It was moved to the museum basement for safety and Orville lived to know the terrible destruction his invention was capable of.

CHAPTER 2

SEEING WITH ELECTRICITY

The very first time John Logie Baird managed to transmit an image that convinced him he had cracked the problem of 'seeing with electricity' there appeared on his tiny screen the grotesque severed head of a ventriloquist's dummy. The intense light required to generate a picture was too much for human subjects who would not, anyway, have wanted to sit for Baird while he fiddled with his ramshackle equipment. So he acquired the dummy with its articulated jaw and staring eyes and nicknamed it Stooky Bill. Baird was a Scot where Stuccie is a Scots term for a plaster cast. Stooky served his purpose: though he was just a dummy, he had features that would only be captured once Baird's technology had achieved a certain degree of sophistication. Up to the time Stooky's head appeared, Baird had not been able to transmit anything more than shadows or images of individual letters.

The contraption that captured images of the dummy's head and transmitted them a few yards bore no resemblance at all to a modern television set. With its whirring disks it looked more like a machine for slicing things up, which is precisely what it threatened to do on occasion, when pieces flew off, spinning with menacing speed. Parts had been cannibalised from a variety of everyday bits of material and machinery. Cardboard from a hatbox and the side of a wooden tea chest were used for the disks, which revolved on spindles made from knitting needles. The motor from an electric fan was hitched to the axle of the disk with a bicycle cog and chain. This was a partly mechanical television, or 'Televisor', as Baird preferred to call it. The electronic part of it transformed the visual impressions produced by the disk, which whirred at 800 revolutions a minute (rpm), into a pulse which was then transmitted to a receiver that was a mirror image of the transmitter. The day this bizarre assemblage of parts miraculously caused Stooky Bill to appear as a recognisable image on Baird's tiny screen was 2 October 1925, a date which is generally (though not universally) agreed was the very first time television really 'worked'.

Certainly Baird liked to recall this breakthrough his 'eureka moment'. In his brief memoir *Television and Me*, which he dictated years later while he was recuperating from one of his many bouts of illness, he recalled: 'The image of the dummy's head formed itself on the screen with what appeared to me to be almost unbelievable clarity. I had got it! I could scarcely believe my eyes and found myself shaking with excitement.' Baird remembered rushing down the stairs from his makeshift studio in London's Soho and grabbing the first person he could find to come upstairs to sit for him. This was twenty-year-old William Toynton, the office boy on the ground floor. Baird placed him in Stooky Bill's position and then dashed next door. To his consternation, when he tried to view Toynton there was no image on his tiny screen. He went to check his transmitting equipment and found that William had recoiled from the intense light and was no longer in position. Baird persuaded him to return with a bribe of half a crown (2s. 6d). Back in front of his miniature screen he shouted: 'William, I can see you! I can see you!' In interviews he gave long after the event, William said Baird had asked him to put his tongue out, which he had done reluctantly: he felt it was rude but was persuaded to overcome his scruples in deference to technological innovation. Baird had wanted to know just how much detail he could capture with the Televisor. And to make absolutely sure that he was not imagining what he saw, Baird changed places with Toynton, who was able to confirm that he could see Baird on the miniature screen just as the inventor had seen him.

Baird's account suggests it was Toynton who was the first person ever to appear on television. Others later claimed to have sat for him before October that year. Baird's gift for invention was not confined to electronics or mechanics: he liked to spin a yarn and his memories were not always entirely reliable. Whether it was Toynton or not does not really matter. It is the date that is significant. As with all inventions, a time had been reached when someone was bound to fulfil the dream of seeing by electricity. The fact that it was achieved by an eccentric Scotsman who lived on a pittance and liked to foster the image of the slightly dotty inventor is not as surprising as it might seem. During the pioneer years of television sophisticated equipment simply did not exist and a gifted and determined amateur, as Baird was, had as much chance as anyone of making it work with parts cannibalised from bicycles and radio sets and the electric telegraph.

In the history of invention and innovation Baird's early success is a classic example of the amateur enjoying a 'eureka moment'. However, the bits and pieces the tinkering inventor had to work with had a venerable history. The creation of television had many beginnings, one of the earliest and most significant of which was a discovery made as long ago as 1817 in a mineral mine close to Great Copper Mountain in southern Sweden.

* * *

The Swedish chemist and mineralogist Jacob Berzelius travelled a great deal within his native country and throughout Europe, and wherever he went he carried with him his most cherished piece of scientific equipment: a blowpipe. He demonstrated to some of his distinguished hosts the quite brilliant analytical possibilities this simple tool made possible. In 1822, at the spa town of Eger in Hungary, he met Goethe, who at first ignored him but, recognising the expertise Berzelius had in identifying minerals, showed him some rocks he had collected locally. 'They were arranged in a special room on a large table and were rather extensive, but very few specimens had been worth bringing home,' Berzelius recalled in his *Autobiographical Notes*:

> In one case I was of a different opinion than he as to the name of the mineral which was shown, and since Goethe would not believe me, I proposed to him to settle the question with the blowpipe test. He replied that he had little knowledge of the use of this instrument, but would, however, gladly see the experiment with it. I occupied a room in the same inn and fetched my apparatus, which I always carried with me on my journeys. Goethe was so struck with the certain results obtained with it that he insisted upon my testing with the blowpipe a number of things he had collected. . . . When I showed him how easily titanium is demonstrated by a beautiful reaction he lamented, feeling that his years now prevented him from perfecting himself in the use of the blowpipe.

Although Berzelius was prone to whip out his blowpipe at the table of his hosts to check the true nature of the salt, he felt restrained when he was entertained in London in 1812 by one of the most distinguished scientists of the day, Sir Humphry Davy. The two had corresponded regularly – Berzelius is said to have written about 100,000 letters in his lifetime – and had much to talk about. But Davy's household took him aback: he left his card three times with a French butler before he was finally ushered in to share breakfast with Davy in a lavishly furnished and decorated living room. Though born into a humble family, Davy had married a wealthy widow and lived in style. In contrast, Berzelius was not then married and, in fact, remained single until he was fifty-six years old, devoting his life to his work. Of Sir Humphry's laboratory, where Berzelius felt more at ease, he wrote: 'When I saw the collection of broken vessels, of melted slagged retorts, those tables covered with marks from acids and caustic alkalis and ring after ring left by vessels whose contents had boiled over . . . I arrived at the happy conviction, which previously had been only a guess, that a tidy laboratory is the sign of a lazy chemist.'

Although Berzelius did make use of a variety of pieces of chemical equipment he never abandoned the blowpipe and was keen to promote its use outside Sweden: it had such simplicity and was so effective when used with knowledge and skill.

An English handbook published in 1825 entitled *Instructions for the use of the blowpipe and chemical tests with Additions and Observations derived from the original recent publication of Prof. Berzilius* has the following:

> The blow-pipe is a most valuable little instrument to the mineralogist, as its effects are striking, rapid and well characterised and pass immediately under the eye of the operator. Small difficulty generally attends the first attempts to use it, but with a little perseverance the habit will soon be acquired; perhaps no caution can be more essential than that of *not working too hard* as the most efficacious flame is produced by a regular stream of air, while the act of blowing with more force only has the effect of fatiguing the muscles of the cheeks, oppressing the chest and at the same time renders the flame unsteady.

Just a few inches long, with a curled tip, the blowpipe was, in effect, the bellows of a miniature blast furnace and had a history going back to antiquity. Reed blowpipes were made by the Egyptians long before the Christian era, and they were widely used in the Middle Ages as a tool of craft workers, such as jewellers, who needed intense heat to solder metals together. In Sweden it was first an instrument used to melt down ores for smelting before it was developed as a vital piece of equipment in the chemical analysis of rock. The eighteenth-century mineralogist Axel Fredrik Cronstedt used the blowpipe to begin a systematic classification of minerals which he produced as a thesis in 1758. A pupil of Cronstedt, who was in charge of a mineral mine in Falun, Johan Gottlieb Gahn, continued the practice and the classification of minerals and, as an elderly man, it was he who taught Berzelius how to use the blowpipe. Berzelius recalled his visits to the Falun mine and Gahn's examination of the rocks found there: 'It was surprising to see the speed and accuracy with which he could identify minerals and how traces of the metals, which otherwise would certainly have escaped the eyes of the analyst, could be detected and identified . . . I learned his method of handling the blowpipe with which he had acquired unusual skills.'

Berzelius had had a tough life. Born in 1779, Jacob was just four years old when his father, a clergyman, died. His mother then married another clergyman, who taught him to read and write. When he was nine years old his mother died and he was cared for by various members of his adopted family before he settled with an uncle who already had seven children and an alcoholic wife. As a boy he spent much of his time hunting birds and collecting insects in the Swedish countryside. Always a bright youth, he won a place at medical college and qualified as a doctor. He came relatively late to chemical analysis and the blowpipe, and joined a distinguished coterie of Swedish mineralogists.

From the first half of the eighteenth century, Swedish mineralogists and chemists discovered a host of elements that have played an essential part in the development of all kinds of modern technology. To name just a few, there is Cronstadt's identification of nickel, Carl Wihelm Scheele's discovery of tungsten, Gahn's discovery of manganese and the discovery of lithium by Johan August Arfvedson. But, in the history of television, none was more significant than the discovery made by Berzelius of an element that he named *selenium*. As so often in the history of innovation, chance played a part in the discovery.

In 1800 a factory had been established next to Gripsholm Castle in Sweden to produce alcohol for the production of acetic acid, which was in demand for the manufacture of white lead paint. The project failed and in 1816 the factory was put up for auction. Some entrepreneurs bought it and looked for chemists who might turn the business around. They brought in Gahn who, in turn, asked Berzelius to join him as a consultant.

Part of the process involved making sulphuric acid from iron pyrites. The former factory manager had rejected the ore that came from the Falun mine because it left a reddish sludge in the lead chamber. Gahn and Berzilius, however, became intrigued to know what this sludge might be and began to analyse Falun pyrites. They thought at first that they had found traces of the element tellurium as it burned with a bright blue flame and gave off a strong smell of horseradish when examined with the blowpipe. But this element had never before been found at Falun where Gahn and others had been researching for many years. After a meticulous process of elimination during which he broke the rock down into its constituent parts, Berzelius concluded that he must have isolated a new element. When it was established that this was in fact the case, it was necessary to find a name for it. As it resembled tellurium, which took its name from *tellus*, the Latin for earth, Berzelius decided to name this related element after the Greek for the moon, *selene*. And so it was that selenium became known to science and available for experimentation of many kinds. Though the element rarely occurs in a pristine form it is easily derived from a number of industrial processes, such as the smelting of copper.

Berzelius had experimented with electrochemistry but he died long before the extraordinary property of the element he had discovered became known. In 1835 he had married the daughter of a friend with whom, as a bachelor, he had spent the previous twenty Christmases. The bride, Elizabeth Poppius, was just twenty-four years old, barely half her husband's age, but she took an interest in his work and by all accounts they had a happy marriage. There were no children, but married life tempered the relentless regime Berzelius had followed for years, often writing from 6.30 a.m. until 10 p.m. As a wedding gift, he was made a baron and towards the end of his life was showered with decorations by many countries; he wrote that he 'had more of these things than can be hung around the neck and pinned to the coat of a scientist even on the great occasions without making him ridiculous'. He died on 7 August 1848 and was buried in Stockholm.

Berzelius left behind formulae for ninety different selenium compounds. After much difficulty he had managed to isolate some pure selenium crystals so that he could determine its atomic weight. But despite all the analytic work he did Berzelius never knew that selenium, in several of its forms, could act as a kind of photoelectric cell. This discovery came about by chance a quarter of a century after his death.

* * *

One characteristic of selenium that Berzelius had noticed was that it was a poor conductor of electricity. Subsequent research showed that its conductivity varied according to which of the various forms it took. In one form it could be used industrially, when a very high resistance to electricity was required. And it was in this incarnation that the first of its remarkable properties was disclosed. Once again, there was an element of chance in the discovery. It began with the development of a new form of insulation for underwater telegraph cables.

A Scottish doctor, William Montgomerie, working as an assistant surgeon for the East India Company Army in Singapore, thought that a rubber-like material used in Malaysia to make handles for machetes or *parangs* might be useful for fashioning some kinds of surgical instrument. It was known as gutta-percha and was made by cooking a latex liquid tapped from several species of indigenous tree. In 1843 Montgomerie, who had an interest in agriculture as well as in medicine, arranged for samples of the gum to be sent to the Royal Society of Arts in London. It took some time to find the formula to turn it into a useful substance but when this had been discovered there was a gutta-percha mania. It was the natural forerunner of plastic and could be put to an astonishing range of uses.

A number of companies were formed to exploit it, one of the earliest of which was the Gutta Percha Company of London, founded in 1845. A young man from the East Anglian town of Great Yarmouth, Willoughby Smith, joined the company three years later and began a career in which, strange as it may seem, he reported a discovery which triggered a feverish excitement in the world of telegraphy and led some to believe that the problem of 'seeing at a distance' had been solved. Smith began in the gutta-percha business just at the time when it was discovered that it was not just useful for making golf balls or a range of oddments. The first long-distance underwater telegraph cables were being laid and it was discovered that gutta-percha was a much more durable insulator than any other material. Smith became involved in cable laying, working on the first abortive attempt to run a seabed telegraph across the English Channel in 1850 and a successful attempt the following year. In time the Gutta Percha Company merged with a telegraph company to become the Telegraph Construction and Maintenance Company (Telcon) and Smith became an expert in the laying of submarine cables.

By 1873 Smith had retreated to a shore job after a bout of illness and was looking for a low conductor of electricity for use in the testing of telegraph cables. In a letter he wrote to the vice president of the Society of Telegraph Engineers he reported a discovery which was to link Berzelius's discovery of 1817 and Baird's 'eureka moment' in 1925: 'I was induced to experiment with bars of selenium – a known metal of very high resistance. I obtained several bars. . . . Each bar was hermetically sealed in a glass tube, and a platinum wire projected from each end for the purpose of connection.' The results were not satisfactory, he wrote, because different operators got different results when they tested the conductivity of the bars. 'When investigating the cause of such great differences in the resistance of the bars, it was found that the resistance altered materially according to the intensity of light to which they were subjected.' Smith signed off with an apology that he could not attend the next meeting of the Society but offered to deliver a paper in future giving more detail of his observations on selenium if his communication was 'of sufficient interest'.

It was the observations of Smith's chief assistant, Joseph May, which had revealed that the selenium cell was changing its resistance according to the intensity of the light falling on it. Berzelius had not spotted this chameleon-like property of the element he had identified and neither had a number of scientists who had examined it in the half-century since it had first been discovered. A reasonably sensitive instrument for measuring weak electrical currents accurately was a prerequisite for May's observation, and none was available to technicians until the second half of the nineteenth century.

When Willoughby Smith's account of the strange property of selenium was made known there was a great flurry of interest in this chemical element and many uses were found for it. There is nothing simple about selenium. It can take many different forms, each of which has different properties. Selenium's variable conductivity under different intensities of light was employed in many ingenious ways. As its resistance varied between the hours of darkness and daylight a selenium cell could be used to turn public lighting on and off with a circuit designed to trip a switch according to the strength of light falling on it. Of all the many exciting possibilities selenium promised, the most anticipated was the prospect it held out of solving the problem of seeing by electricity. Long before the success achieved by Baird and others in the 1920s there had been many ingenious attempts to create television with this new element as the essential component, as it could convert visual images into electric currents which could then be turned back into visual imagery by a receiver. However, if any of them were to be made to work they would be dependent on a peculiarity of human visual perception that had intrigued enquiring minds since antiquity.

* * *

In the 1820s there was a vogue for the creation of simple toys that would be both entertaining and instructive. They were the invention, invariably, of scientific men who were interested in all kinds of optical illusion. The 'wonder turner' or thaumatrope (meaning 'magical motion') first appeared in England 1825, its invention attributed to a number of people, including the English physician John Ayrton Paris. It is so simple you could make one yourself if you put this book down for a moment. All you need is a disk made of card, say two inches in diameter, a pen and two bits of string. On one side of the disk you draw a birdcage and on the other a bird. On either side of the disk you punch holes, to each of which you attach a piece of string. Pull the opposing strings tight and spin the disk. When a critical speed is reached the bird will appear to be inside the cage. It is a novel optical illusion, but by no means the first to interest inquisitive minds.

In his epic work, *The Nature of Things*, the Roman poet Lucretius wrote: 'And when the winds carry thin clouds across the sky at night, then brilliant stars seem to glide in the opposite direction against the clouds, moving high above them on a path very different from the one they really travel.' In the same passage he noted: 'Then, when a spirited horse is stuck fast in the middle of some river and we look down at the rushing waters of the stream, some force appears to carry the horse's body, which is not moving, sideways to the current, to be driving it rapidly upstream.' There are so many familiar optical illusions that we tend to think nothing of them: such as, for example, the fact that if we rapidly twirl a lighted torch or sparkler in circles it appears to draw a continuous line in the darkness.

It was proposed by Isaac Newton, among others, that the effect was due to what became known as the 'persistence of vision'. The speed of rotation is critical for the illusion to take effect, suggesting that it is a consequence of some mechanism whereby a visual image is retained for a fraction of time and will blend with a new image if this appears almost instantly. Newton suggested that the retention of an image lasted less than a second. An attempt to make a more accurate estimate was devised by the brilliant mathematician and scientist Patrick d'Arcy.

Born in 1725 in the west of Ireland to an Irish mother and French father, D'Arcy lived most of his life in France where, as a fourteen-year-old, he had been sent by his parents to escape religious persecution. He was for a time a French Army officer and was made a chevalier, the French equivalent of a count. Although research into vision was not his main interest he was intrigued enough by Newton's guess at the duration of visual persistence to devise, in 1765, a series of experiments to give the apparent time lapse a definite value.

The setting for D'Arcy's investigation was a darkened building in which he constructed a revolving cross, the speed of which he could measure. To one arm of the cross he attached a burning coal and set it spinning. As it gathered speed he noted the point at which the coal appeared to draw a continuous line in the dark. He spun it back and forth, sometimes noting the critical minimum speed for the illusion to take effect as the cross lost momentum. He changed the position of the coal on the cross; viewed it through a telescope; squinted at it through a pinhole; and changed the distance from which he viewed the spinning coal. The results all seemed to be the same, which suggested that the retention of an image lasts 130 milliseconds.

This optical effect of the whirring piece of coal was, it seemed, the result of the same mechanism whereby the bird in the twirling thaumatrope appeared to be in the cage, and it was exploited by scientists, who devised gadgets to illustrate its many possible manifestations. Some of these became familiar toys in the Victorian nursery and can still amuse young children by their magical, cartoon-like effect. One of the most popular was the zoetrope in which a sequence of images viewed through vertical slits in a revolving drum give the impression of continuous movement: a horse jumping, or a juggler juggling, or a couple dancing. A curious aspect of this effect is that when the revolving drum reaches a critical speed the strips that divide the rapidly revealed images become invisible. There was nothing essentially new in this discovery: Leonardo da Vinci had remarked on the fact that perforated cards used by women to gather threads could be 'seen through' when spun at a certain speed.

What was new about the philosophical toys of the nineteenth century was that they pointed the way to the exploitation of visual illusions by new technologies. The zoetrope was a simple form of animation, the drum spun by hand or by a clockwork motor. The illusion of movement was achieved using images drawn by hand and the zoetrope continued to be a popular form of entertainment long after the appearance of the photograph which, in its first, primitive forms, could not capture images of rapid movement. Strangely, however, though these gadgets and the development of instantaneous cameras and film appeared to point the way to the quest of 'seeing at a distance', a critical nineteenth-century breakthrough in the evolution of the idea of television had more to do with sound than with sight.

* * *

In 1876 news that a new instrument called the telephone enabled people to speak to each other though they might be miles apart was greeted as one of the wonders of the age. Within a year or two of the telephone's appearance, newspapers and magazines began to publish futuristic cartoons in which people could not only speak to each other at a distance, they could see each other too. In the popular imagination, it seemed, sound and vision were not that different and among scientists there was a growing conviction that if speech could be turned into an electric current, and that current could be turned back into speech, then perhaps the same principle could apply to vision. Some science fiction writers anticipated the transmission of entertainment or events in the real world. Television was 'imagined' long before it became a practical possibility, and the telephone was the device that inspired much futuristic speculation.

A cartoon published in the satirical magazine *Punch* in December 1879 illustrates a telephonoscope which 'emits light as well as sound', its creation attributed to the celebrated American inventor Thomas Edison. A couple ready for bed sit before the fire clutching their telephone speakers and on a screen above the mantelpiece watch their children playing tennis – in Australia! The father speaks to his daughter to ask who the delightful young lady is on the other side of the net. A little later, in 1884, the French illustrator and cartoonist Albert Robida published a book anticipating the twentieth century and the 'conquest of the regions of the air'. In one illustration a family recoil in anxiety as they watch a live scene from a battlefield. In another a family watch a scene from the opera, and there is a man reclining in a *cabine telephonoscopique* with the caption saying that these are to be found on every street corner in the 'better neighbourhoods throughout France'.

Although this riot of imagination came well before the invention of wireless, the world was linked by telegraph cables by the 1870s so the prospect of long-distance transmission of pictures by cables like those which connected telephones was not entirely outlandish. It was rumoured, in fact, that Alexander Graham Bell, the young Scotsman who had emigrated with his family to Canada and then moved to

America, was on the verge of announcing that he could send moving pictures by wire. Bell had patented the telephone in 1876 (for the full story see Chapter 5 on mobile phones) and in August 1880 had presented a paper in Boston to the American Association for the Advancement of Science entitled: 'On the production and reproduction of sound by light'. He, like so many others, had been inspired by the discovery of the photosensitivity of selenium, and he gave due credit to Jacob Berzelius in his address for discovering the element by his meticulous blowpipe analysis.

What Bell was proposing, however, was not television but a kind of wireless transmission of speech. Sounds would be converted into light, which would then beam a charge across to a receiver so that people could speak to each other without the need for a cable connection. Bell got what he called his photophone to work but it was soon redundant (see Chapter 5 Hard Cell for the story).

Turning the sound waves of the human voice into an electric impulse which could then be turned back into speech, wonderful though it was, proved to be a good deal simpler than the task of creating distant vision. This was because turning a visual image of any kind into a set of pulses required hundreds, if not thousands, of light-sensitive cells to 'map' just one still picture. To capture a continuously moving scene, as imagined by the cartoonists and futurists, was thought to be impossible by many who puzzled over the problem. And in the 1880s it was out of the question because vital technologies had not yet appeared. There was one key concept, however, which was understood: that the only way to capture an image was to scan it.

By the 1880s rudimentary scanning of still images had been achieved and the techniques of what became known as telephotography progressed rapidly from the ability to copy and send simple shapes by wire, to whole images and eventually photographs. The development of photography itself, particularly the advances made in the speed of film and camera shutters, seemed to hold out the promise that objects moving at speed could be snapped as if frozen in time. In practical terms, the quest for what was known as 'instantaneous photography' had nothing to offer those seeking to solve the problem of distant vision. Nor did it lead directly to the development of moving film. But the discovery that there were startling limitations to human vision which could be compensated for and corrected by photography fired the imagination of the public, who would flock to see performances of moving images that were really not much more gripping than the silhouettes flickering by in the nursery room zoetrope. This fascination with the same old images made exciting by the *way* which they were presented survived right up to the early days of television.

* * *

In July 1860 an Englishman calling himself E.J. Muygridge, then thirty years old, left behind in the hands of a younger brother a bookselling business he had established in San Francisco and began a journey back to his home town of Kingston upon Thames in England. He planned to go overland to New York and ship from there across the Atlantic. For the first part of his journey he chose to travel on a recently established fast route to St Louis run by the Butterfield Overland Mail Company. Three weeks into the journey, the brakes of the stagecoach failed as it ran down a mountainous road in the woodlands of central Texas and the six wild mustangs hauling it galloped off.

When Muygridge regained consciousness nine days later he was lying in bed in a place called Fort Smith, 180 miles from the scene of the accident. He learned that he was lucky to be alive: the stage had hit a tree and disintegrated. Two other passengers had been killed. He later told the *San Francisco Daily Evening Bulletin*: 'I found a scar on my head. I had a double vision – saw two objects at once; had no sense of smell or taste; also had confused ideas. These acute symptoms continued three months. I was under treatment about a year.' When he heard of the circumstances of the crash that had nearly killed him he decided to sue the Butterfield Overland Mail Company, seeking $10,000 in damages. This he did after returning from London to New York in 1861, settling for $5,000.

Muygridge had left San Francisco in 1860 a bookseller. He returned there in 1867 as a photographer, calling himself Edward Muybridge. It has been said that he was keen to make a name for himself – and he invented more than one. Born Edward James Muggeridge, on his first trip to America he had doctored his surname to give it a quaint Anglo-Saxon timbre. Then Muygridge became Muybridge when he turned photographer, though here again, his professional name was Helios. In the 1880s he elongated his first name to Eadweard. There is no record of how or where Muybridge learned to take and develop photographs, but it must have been in London in the 1860s when this new technology was very exciting and rapidly evolving. And he had his $5,000 to spend. His absence from America between 1861 and 1867 is likely to have been because of the outbreak of the Civil War between the Confederacy and the Union, which lasted from 1861 to 1865.

Setting himself up again in San Francisco, Muybridge first gained a reputation for his giant wet-plate photographic portraits of the spectacular scenery of Yosemite in eastern California and other wild landscapes in the American West. The equipment he used was cumbersome, requiring on occasion a team of porters. It was worth the trek to the wilderness, however, as it was rapidly being opened up and settled as the Native peoples were driven from their ancestral hunting grounds. A landmark in this conquest of the continent was the completion of the Union Pacific Railroad on 10 May 1869; this joined the newly forged United States from the East Coast to San Francisco in the West. At a ceremony in Utah one of the principal promoters and financiers of the railway, Leland Stanford, drove a golden spike into the track.

Stanford and the other railroad magnates made fortunes and flaunted their wealth in the nouveau riche culture of San Francisco and the Californian coast. Hundreds of horses had been used in the laying of the Union Pacific Railroad to haul the iron tracks a few yards at a time. In both the towns and the countryside horses remained the most important means of transport pulling streetcars (trams) and farmers' wagons. And there was no grander form of conspicuous consumption for a railroad millionaire than a stud of fine trotting horses. How it came about that Muybridge was chosen to make history by photographing one of Stanford's prized trotters called Occident is clouded in mystery. Here is an account from the Preface of Muybridge's book *Animals in Motion*, published in 1899:

> In the Spring of the year 1872, while the author was directing the photographic surveys of the United States Government on the Pacific Coast, there was revived in the City of San Francisco a controversy in regard to animal locomotion, which we may infer, on the authority of Plato, was warmly argued by the ancient Egyptians, and which probably had its origins in the studio of the primitive artist when he submitted to a group of critical friends his first etching of a mammoth crushing through the forest, or a reindeer grazing on the plains.
>
> In this modern instance, the principal subject of dispute was the possibility of a horse, while trotting – even at the height of its speed – having all four of his feet, at any portion of his stride, simultaneously free from contact with the ground.

Although there was great interest at the time in what had become to be called 'instantaneous photography', fast film had not yet been invented and Muybridge had to work with cumbersome wet-plate cameras and a veritable laboratory of chemicals and equipment to develop his photographs. He wrote:

> In those days the rapid dry process – by the use of which such an operation is now easily accomplished – had not been discovered. Every photographer was, in a great measure, his own chemist: he prepared his own dipping baths, made his own collodion, coated and developed his own plates, and frequently manufactured the chemicals necessary for his work.

It is not entirely clear whether it was in 1872 or 1873 that Muybridge first got the negative he was looking for while filming Stanford's horse

Occident at the racecourse in Sacramento as it trotted across a specially prepared outdoor 'studio' hung with white cloths. Certainly by 1873 there were negatives that gave a conclusive answer to the old question about the gait of the trotting horse: there was a shot of Occident with all four hoofs off the ground. However, this was not the end of the matter but the beginning of a long fascination Muybridge had with the photographing of animals and people in motion. With Stanford's encouragement and backing he devised more and more sophisticated ways in which to take shots in sequence as a horse or some other animal moved across a backcloth. The cameras were lined up in ranks and the shutters primed to 'click' when triggered by a mechanical or electrical mechanism.

The experiments continued at first at the Sacramento racecourse and then moved to a new stud created by Stanford at Palo Alto; this later became the site of Stanford University, which the railway magnate endowed.

One of the extraordinary episodes in Muybridge's epic story is that during the period that he was pioneering 'instantaneous photography' he committed murder, was tried and, despite his admitted guilt, acquitted. He had married a young divorcee in 1871 and a son was born to them in 1874. A letter Muybridge found written by his wife to an English drama critic and rake, Major Harry Larkyns, who had been her constant companion while her husband was away, indicated that the boy was not his but Harry's. Muybridge sought Larkyns out, pulled a gun and shot him through the heart. The plea of insanity proffered by the defence was rejected, but the jury let him off anyway, and the drama appears to have had little or no impact on the onward march of photographic innovation.

When, in 1878, Muybridge's photographs were published in the magazine *Nature*, and its French edition *La Nature*, his fame spread and he was drawn into the vanguard of the world of international scientific and technological experimentation. A French physicist, Etienne-Jules Marey, who had been impressed with Muybridge's photographs, contacted him and they began a collaboration. Marey, who had made his own studies of movement, had a special interest in the flight of birds, which he thought Muybridge might be able to illustrate using his fast cameras. Muybridge later wrote that it was the Frenchman Marey's work on animal movement that inspired Leland Stanford to explore the possibility of photographing trotting horses. Whereas Muybridge used his photographs to create for public exhibition what he called a zoopraxiscope which projected animated images on to a screen, Marey developed a camera gun that could snap pictures at 60 frames a second. Both were forerunners of cinema, which arrived with the celluloid film invented by the American George Eastman.

The very rapid development of the first silent movies in the 1890s took place, it seems, quite independently of interest in the creation of 'distant vision'. Yet a step towards the creation of the earliest forms of television was taken at this time, and though there appears to be no direct connection with the whirring disks of the various forms of the zoopraxiscope and similar mechanisms for animating still images, it is difficult to believe they were not an inspiration.

* * *

In January 1884 a 23-year-old German student, Paul Nipkow, applied for the first ever patent that envisaged a workable television system. What train of thought inspired his scheme remains something of a mystery. Interviewed in the *New York Times* in 1933 he gave the following account:

> One raw winter evening I was cheered by receiving from the post-office the loan of a genuine Bell telephone for two hours. I lived in one room, which served as a living room, sleeping chamber, laboratory and workshop. The remarkable simplicity of the telephone astounded me. It gave me an idea and I constructed a microphone, using nails. It was successful in transmitting noises and words from one attic to another. This experience is what started me thinking about the problem of television.

Nipkow makes no mention of any of those popular nineteenth-century philosophical toys as an influence and it is not at all clear why the mechanism of the telephone should have suggested to him what he called an 'electric telescope'. The component that was entirely original was a perforated disk which, in one revolution, scanned an image and in doing so threw light of various strengths, according to the composition of the image, on to a selenium cell, which in turn sent electrical impulses to a receiver that transformed the variable electric current back into an image with an identical and synchronised receiving disk.

'It was the general idea of television,' Nipkow told the *New York Times*. 'The mental experiment was a complete success. The ideas of the invention were automatically at hand – as all everyday ideas are. How sure I was of having made a great discovery may be seen from the fact that, despite serious financial difficulties, I did not hesitate to spend the money needed to apply for a patent.' He then added: 'Now, however, my pen hesitates. Did I at the time think about the scope and future of the "electric telescope" alias television? Hardly! We must remember that in those days the use of the telephone was only in its first stages. . . . No; my thoughts and worries during the next decades were devoted to my professional work, which was the practical development of the system of making railroad traffic safe. Only occasionally was I able to give any time to my first love – television.'

Nipkow did not attempt to construct the television system he had envisaged. Had he done so it would not have worked, because the electric current from the selenium cell would have been too weak an impulse to transmit the image. There were many alternative proposals for the use of selenium cells at that time which might have worked, although they were otherwise quite impracticable. There was, for example, the model described by the extraordinarily inventive double act generally known as Ayrton and Perry. Born in 1847, William Ayrton was a brilliant student of mathematics at University College London who first worked in British India on improvements to the telegraph system. In 1873 he moved to Japan as a professor of physics and telegraphy in Tokyo where he struggled to turn samurai warriors into telegraph engineers. Faced with sword-waving students he armed himself with a pistol to bring order to his laboratory. In 1875 he was joined in Tokyo by John Perry, who became Professor of Mechanical Engineering, and the two began an intensive experimental collaboration. Both returned to London, Ayrton in 1879 and Perry in 1881, and continued to work together on various projects. One of these was a system first proposed in 1880: of transmitting images using a mass of selenium cells each of which would be linked by cable to a lamp at the receiver, the light and shade of the original image governing the brightness of the bulbs. The English lawyer turned inventor, Shelford Bidwell, who had grappled with the problem of distant vision for some years, dismissed Ayrton and Perry's concept as totally impracticable. He calculated that it would have 90,000 working parts and would cost £1,250,000 (around £100 million at 2013 values). Ayrton and Perry made no attempt to build their television, startling Londoners instead with their electrically driven tricycle which joined the horse-drawn traffic in 1882.

There were many experimenters in the 1880s who devised what they believed would be practicable schemes for seeing with electricity, all of them involving the use of selenium cells and some system of scanning images. None worked, and one of the reasons for this was that there was a component missing, a device which would amplify the weak electric impulses produced by the action of light on selenium. The breakthrough came from a quite unexpected branch of electricity: an enigma that became known as the 'Edison effect'.

* * *

At more or less the same time as the young Nipkow patented his electric telescope, a research team in America working on the newly devised filament electric light bulb were puzzled by a recurring problem. The first commercial incandescent light bulbs had been produced more or less simultaneously by the chemist Joseph Swan in England and Thomas Edison's laboratory at Menlo Park in New Jersey. A great deal of experimentation went on in the search for the most effective material for the glowing filament. Edison and his men for a time used carbonised bamboo, but they found that the vacuum bulbs enclosing the filament had a tendency to be darkened by a deposit on the inside of the glass. It was discovered that the blackening could be reduced by inserting an extra metallic plate; this appeared to absorb the carbon emissions which had been spoiling the bulb.

While puzzling over this, an assistant at Menlo Park connected the extra metal plate to an electric meter to discover what current was flowing in it. He found that the result was different according to the side the meter was connected to: on the positive side a current flowed, whereas on the negative side there was practically nothing. This discovery of the one-way flow of current was later dubbed the 'Edison effect' by William Preece, Chief Engineer of the British Post Office, who had been intrigued by it. Edison, characteristically, patented the discovery in the belief that it might be useful for something, some day. He was right, though he had little idea how or why, and it took years of experimentation before the Edison effect was turned into a useful tool.

Nobody spent more time puzzling over the reasons of the darkening of the bulb than the English physicist Ambrose Fleming, a professor at University College London. He was a small, combative mustachioed figure who was hard of hearing. When the young Guglielmo Marconi was planning to send a wireless signal across the Atlantic, Fleming was his chief adviser. Fleming also worked for a time as a consultant to Edison's British company. In 1903 Marconi was embarrassed when a rival ambushed a Fleming lecture in which he was illustrating the wonders of the new wireless telegraphy by sending a message which anyone in the audience who could interpret Morse Code would translate as: 'There was a young man from Italy who diddled the public prettily.' Fleming did not pick this up because of his deafness but he was furious when he learned about it the following day and protested in a letter to *The Times*, effectively making an ass of himself. He was mortified when his consultancy with the Marconi Company was terminated and he was keen to win back the lucrative position.

It has been suggested that it was this determination to show that he still had something to offer that concentrated Fleming's mind on the possible uses of the Edison effect. He told the story much later that he had one day had a 'happy thought' that the one-way flow of electricity in an adapted light bulb might enable it to act as a receiver of wireless signals. There is a story of him scurrying along to the Patent Office in 1904, keen to get his invention protected before someone else thought of it. Fleming contacted Marconi to tell him he might have chanced upon a wireless receiver that was an improvement on the favoured Maggie, or magnetic detector, then in use on ocean liners. Because the current in the adapted light bulbs ran only one way they became known as 'valves' and after a good deal of experimentation they were found to work, though no better than the magnetic detector.

However, Fleming's inspiration spawned a rapid development of all kinds of 'valves' as experimenters tinkered with the internal workings of the bulbs. In America the inventor and rival of Marconi, Lee de Forest, added a grid to Fleming's valve and found that it not only acted as a better receiver, it amplified signals as well. When Fleming read in a technical journal in 1906 of de Forest's 'audion' he was enraged and immediately challenged the American's claim to the invention. But the audion was a distinct improvement on Fleming's valve and it was soon modified and enhanced by others who understood the workings of it better than de Forest himself.

In effect, a problem with the incandescent light bulb gave rise over a period of more than twenty years to a wireless valve that could amplify sound. And it was this invention that provided the missing component in the development of television. Valves could be used to boost the weak electrical signals emitted by a photoelectric cell. It was time to string together the discoveries of centuries and to make the first realistic attempt to 'see by electricity'. With the development of wireless and, from the 1920s, broadcast radio, the possibility existed that visual images could be sent without cables, perhaps even across the oceans. It was a thrilling prospect, yet the task of making it work was, by and large, left to amateurs such as Baird while the newly created broadcast and telephone companies concentrated on exploiting the existing technologies. The scientific understanding of electrical equipment, which was becoming infinitely more sophisticated, went its own sweet way until the time was ripe for the exploitation of its astonishing discoveries.

* * *

In his unreliable memoir dictated in 1941, John Logie Baird recalled the moment he conceived of his world-beating television system. It was 1922 and he had joined his old friend Guy Robertson in the seaside town of Hastings on the south coast of England. Baird said he had gone there for his health as he was suffering from one of the persistent colds that dogged him all his life. Hastings had a number of radio hams, amateur enthusiasts who formed the Hastings Wireless Society in 1924. It is possible that it was their presence in the town and along the south coast which prompted Baird to have a go at television. He would not have got very far without them. This is how Baird recalled breaking the news to his fellow lodger Guy, whose nickname was Mephy:

> I went for a long walk over the cliffs to Fairlight Glen, and my mind went back to my early work on television. Might there not be something in it now? My difficulty then had been to find a means of amplifying the infinitesimally small current from the selenium cell. Such an amplifier was now available, thanks to Fleming and de Forest. Why not try again? The more I thought about it the easier it seemed. I thought out a complete system and returned to Walton Crescent [his lodgings] filled with an influx of new life and hope. Over the raisin pudding I broke the news to Mephy: 'Well Sir, you will be pleased to hear that I have invented a means of seeing by wireless.' 'Oh', said Mephy, 'I hope that doesn't mean you are going to become one of those wireless nitwits.'

Radio broadcasting was still a novelty in 1922 and the 'wireless nitwits' were the amateur radio hams who formed the first audience for the few programmes aired under strict government regulations by the Marconi Company in London and two other regional broadcasters. It would be a few years before 'listening in' became a nationwide habit. The British Broadcasting Company Ltd had been formed in 1922, financed by radio manufacturers and the Post Office, becoming in 1927 the British Broadasting Corporation funded by a licence fee with a monopoly on programming. Mephy – short for Mephistopheles, a nickname Robertson acquired as a youth, apparently for his 'dark good looks' – was justified in thinking his friend Baird might be joining a band of hobbyists rather than becoming a communications trailblazer.

Doubtful though he was, like a good friend, he lent Baird a hand and in no time at all they had their first prototype television. It was, Baird recalled, a 'very simple' apparatus. Everything he needed was to hand: a hatbox and some tin plate for his spinning disks, darning needles for spindles, an electric motor, bullseye lenses from bicycle lamps, a neon light, a selenium cell, and a valve amplifier. Though Baird's first arrangement of these parts was ingenious it was made possible by hundreds of years of scientific and experimental endeavour, of which the amplifier valve was the most up-to-date invention. He also relied heavily on the recently acquired expertise and equipment of the pioneer radio amateurs of Hastings, for Baird himself had never been a 'wireless nitwit'.

Baird had a comfortable upbringing in a substantial house called the Lodge in Helensburgh, Scotland. It had been bought by his father, a minister in the West of Scotland Church, with a 'dowry' from his wife, Jessie Inglis. Jessie was orphaned as a young girl and was taken in by an uncle who was one of two Inglis brothers who ran a very successful shipbuilding firm on the Clyde. They built the famous paddle steamer, the *Waverley* as well as *The Maid of the Loch*, which remains a tourist attraction on Loch Lomond with plans to get it steaming again at the time of writing (2013). Baird's older sister, Annie, described in her diaries happy days at the Lodge, with traditional pastimes such as croquet on the lawn as well as up-to-the-minute entertainments including a lecture on South Africa illustrated with a magic lantern, and even an early film.

Born in 1888, Baird was a sickly child cared for by a devoted mother on whom he doted. He showed signs of inventive inquisitiveness as a schoolboy, rigging up a telephone system with friends with, according to his own account, unfortunate consequences when a trailing wire unseated a carriage driver. It seems he knew of the properties of selenium and of the Nipkow disk as a boy and attempted to make a selenium cell, burning his hands in the process. His father urged him to study for the Church, but John preferred science and enrolled at the Glasgow and West of Scotland College of Technology. It was an austere institution for hard-working boys who wanted to better themselves. When he graduated he was accepted at Glasgow University where, by his own account, he enjoyed the decadent pleasures of the wealthier students. In his pioneering days as an inventor Baird played up the idea that he dressed raggedly and lived in miserable lodgings because he was broke. He certainly lacked funding for his television project early on but he did not happily suffer for his cause. Whenever he could, he wined and dined. It was his wealthy cousins, the Inglis brothers, who made sure Baird did not starve. Exactly how much they gave him over the years is not known.

Baird had begun his degree at Glasgow University in 1914 when war broke out. In 1916 conscription was introduced for the first time and Baird dutifully presented himself to the recruiting office. He was turned down on medical grounds, though it appears that he made vain attempts to join up anyway. In the end his war work was at Clyde Valley Electrical Power Company where he was taken on as an assistant engineer. The pay was not bad, but he hated the job and began to dream of making a fortune. In fact what really appears to have motivated Baird at that time was the desire to become rich, somehow. When he was at college he had an article published which was not entirely tongue in cheek. With hindsight, it reads like a manifesto, and Baird stuck to it with grim determination. Giving it the title 'How to make money', Baird wrote under the nom de plume 'H2O':

'Some geniuses put whisky and water, with a little cinnamon and sugar, into medicine bottles, label it "Swamp Root Tonic Laxative – a

Pure Vegetable Extract", and sell it at a shilling a bottle. They make millions. . . . Other geniuses put salt water into bottles and label it "Fruit salts – a healthy mind in a healthy body". They make millions.' After a long list of millionaire charlatans his article continued: 'Thousands upon thousands of d——d fools drudge all their lives in drawing offices. They make from 24s to £3.10s a week.' The snag with the patent medicine route to wealth, according to H2O, was that too many people were trying it. He had a secret method, which he would forward on receipt of a 'postal order to cover postage and clerical expenses'. He closed with the insistent advertising slogan of 'Do it now'.

During this literary period of Baird's youth he had eighteen articles and stories published in the college magazine, some of which reflected his great admiration for the futurist writings of H.G. Wells. This inspired his near-disastrous attempt to use the power station mains to transform carbon into diamonds. The blackout was quickly reversed. While still drawing his wages at the power station he tried first to create a cure for piles, with uncomfortable results, and then marketed what he called Baird's Undersock. His explanation for this inspiration was that his own feet were always cold. The undersock, dusted with borax, fitted in the shoe and kept feet warm. He advertised it on billboards and shortly after left the Clyde Valley Electrical Power Company.

There followed a series of enterprises including jam making in the West Indies, the manufacture of soap, and a last and disastrous attempt in Hastings to create a new kind of luxury footwear. He recalled:

> I decided to try pneumatic shoes, so that people could walk with the same advantage that a car gains from its pneumatic tyres. I got a pair of very large boots and put inside them two partially inflated balloons, and then very carefully inserted my feet, laced up the boots and set off on a short trial run. I walked a hundred yards in a succession of drunken and uncontrollable lurches followed by a few delighted urchins, till the demonstration was brought to an end by one of my tyres bursting.

This clowning, Baird wants us to believe, immediately preceded the inspirational moment in which he thought up a working system of television.

If, as he says in his memoir *Television and Me*, he figured out how to make a working television while walking along a coastal path he must have continued to take an interest in the discoveries and theories of others. There is anecdotal evidence from those who knew him in his power station days that he had tried to build a television system then, while living in lodgings. Perhaps, like many electronics enthusiasts, he had learned the basics at that time. But before radio broadcasting had become established there was very little equipment or technical help available to the amateur: the component parts used in early twentieth-century wireless telegraphy would not have been much use. Just a few years later, in 1922, Baird could find all that he needed by tapping into the expertise and hospitality of the Hastings 'wireless nitwits'.

Perhaps the most important of these amateur experts was a young man called Victor Mills. In a letter to Russell Burns, one of Baird's biographers, Mills recalled the time when Baird began his experimentation in Hastings. We do not know how Baird knew of him but he had clearly been told that Mills was a wireless ham with his own 'shack' and might be able to help. And one day in the winter of 1922–23 Baird knocked on the door of the Mills family home to ask for assistance with his efforts to invent television. There was a specific problem with 'noise' from his selenium cell. Mills was able to offer more than technical help: 'For a while Baird had a room in our house,' Mills recalled, 'where he kept his apparatus. My mother bought two tables on which to do the early experiments, one for Baird and the other for me. . . . I think Baird was a wonderful sticker, he lived television and in spite of a mass of ideas for his work and knowledge of that of others he did not appear to have much real technical or scientific knowledge, but I am pleased to have worked with him.' Many other wireless enthusiasts and engineers in Hastings recalled helping Baird find parts for his equipment.

Baird had got everything from Mills for nothing, including the vital amplifier, but he had to find his own premises and was soon out of pocket. He put a classified advertisement in *The Times* in June 1923 which read: 'Seeing by wireless – inventor of an apparatus wishes to hear from someone who will assist (not financially) in making working models.' He said later he thought it best not to ask for money directly but it seems the readers of the paper took him literally and gave him nothing. Although he had made some money selling soap, Baird had limited resources and could not have continued working on television without help from his wealthy cousins.

Exactly what Baird was up to in 1923–24, or where he was, it is impossible to determine as he disappeared from Hastings for a while and his own recollection was either misremembered or deliberately obscure. But he began to give demonstrations of his crude system, sometimes simply showing radio hams that they could pick up the signal of his transmitted image, but no image. If there was something on his tiny screen it was just a shape or a shadow. He was desperate for publicity that might attract financial backers but always in danger of exaggerating what he could show. The writer Le Queux, a keen radio ham who took an interest in Baird, contributed an article to the *Radio Times* in April 1924 with the title: 'Television – a fact'. He was jumping the gun, but it was the kind of publicity Baird needed and he eventually found a backer from the film industry, a man called Will Day. It was Day who found Baird an attic workroom in Frith Street, Soho in a building owned by a friend in the film industry. In October 1924 he moved in, and a Hastings connection gave him a break the following March.

A dashing chap called Tony Bosdari had been at Winchester public school with Gordon Selfridge Jr, son of the American founder of the Oxford Street store. Bosdari, who worked in the record business, was friendly with a neighbour of Baird's in Hastings and when Selfridge was looking for an attraction for the store in the spring of 1925 he thought of the eccentric inventor and his novel apparatus. In the 1920s Selfridges was hugely popular, attracting more than 20 million customers a year and the management was keen to promote anything modern. A few months after it first opened in 1909 the plane in which Blériot made the first flight across the Channel was displayed in the Oxford Street store.

Gordon Selfridge Jr, who worked for his father, called on Baird and offered him £20 a week for three weeks if he would demonstrate television in the store. A short piece in *The Times* on 24 March 1925, headlined 'Television First Public Demonstration', reported that: 'For the first time in history Television was publicly and successfully demonstrated on the stage in the Palm Court at Selfridges last week. A good deal has been written about television, but here, for the first time, this new wonder was shown in a form which proves scientifically that "it can be done".'

Selfridge's publicity brochure began: 'The House of Selfridge has always gone out of its way to encourage other pilgrims on the Road of Progress. And this picturesque apparatus with its cardboard and its bicycle chain is in direct succession to Blériot's gallant monoplane and Shackleton's brave boat – just to mention two of the world's wonders that were first seen at Selfridges.' After the first show, Baird gave three demonstrations a day in the Electrical Department and won admiration for the novelty of his invention though it was evident to everyone that it was, as Selfridges put it, 'absolutely "in the rough"', the pictures 'flickering and defective'. But then Edison's phonograph had at first produced an unintelligible crackle – and now look at the gramophone playing dance music!

And on 2 October 1925 there was already a breakthrough when a clear image of the dummy Stooky Bill appeared on the screen and Baird knew he was closer to producing something saleable. In his memoirs he tells a fabulous story of how he demonstrated his Televisor to a party from the Royal Institution, an august body with an interest in technological advances. He has his distinguished guests in evening dress queuing on the narrow staircase up to his workshop in Frith Street for a chance to see his wonderful invention. He overhears the comment: 'Baird's got it! The rest is merely a matter of £ s d.'

There is no record of who might have said this, or who the gentleman was whose long grey beard was snared in Baird's spinning Nipkow disk. It is quite probable that both are figments of the inventor's lively imagination. Baird gives the date of this occasion as Friday 27 January 1926. A report in *The Times* on 28 January puts it on the previous Tuesday, 26 January. It is not clear if a reporter was there or the piece was based on hearsay. But there is nothing about evening dress, or beards, or scientists amazed to see themselves on the little flickering screen. The headline read: 'The "Televisor", Successful test of new apparatus':

> For the purposes of demonstration a ventriloquist's doll was manipulated as the image to be transmitted, though the human face was also reproduced . . . the visitors were shown recognisable reception of the movements of the dummy and of a person speaking. The image as transmitted was faint and often blurred. . .

In his biography of Baird, subtitled *The Romance and Tragedy of the Pioneer of Television*, the journalist Sydney Moseley, a close friend and business associate, says that following the demonstration in January 1926 newspapers clamoured for news of the Televisor. Again, this seems to be wishful thinking for there is nothing much in the archives. However, Baird did move his 'laboratory' in February, employed an engineer, a Mr B. Clapp, and took the first steps to set up a company, Baird Television. He was well aware he was in a race. In America there were reports of television broadcasts of a kind, one by the giant AT&T. Baird's response was characteristic. Rather than work on improvements to his system to produce a more acceptable image, he settled on a stunt. In 1927 Clapp was despatched to New York in the hope that he could pick up a signal from London and demonstrate transatlantic television. In 1928 a demonstration was given to the press. It worked, but the image was crude. Nevertheless, as yet, Baird was still in the vanguard, and dreaming of riches. The competition had not yet overtaken him.

* * *

In June 1925 Charles Francis Jenkins published privately a little book which gave an account of the many uses he had found for radio at a time when he was running neck and neck with Baird for the first ever public showing of a working television system. Jenkins had a nice 'down home' turn of phrase as he tutored the general public on the wonders of wireless in his *Vision by Radio, Radio Photographs, Radio Photograms*. Had anyone noticed, he wrote:

> the curious fact that a great laboratory, despite its inestimable contribution to science and engineering, has never yet brought forth a great, revolutionary invention which has subsequently started a new industry, like the telegraph, the telephone and telescope; motion picture, typecasting and talking machines; typewriter, bicycle and locomotive; automobile, flying machine and radio vision. It has always been a poor man to first see these things, and as a rule the bigger the vision the poorer the man. And, do you know, that is right comforting, too: for I sometimes think that perhaps I myself may yet do something worth while if I only stay poor enough, long enough.

When he wrote that, Jenkins was, like Baird, an independent inventor. However, unlike his rival, he already had a notable innovation to his name. He was from a Quaker family that moved when he was an infant from Dayton, Ohio a little further west to Richmond, Indiana where they had a farm. Born on 22 August 1867, he attended a local school and then Earlham College, leaving when he was nineteen to explore the American continent, working in the north-west and down in Mexico, returning home each year. He took a course as a stenographer and passed a civil service exam, which qualified him to work as a clerk. In 1890 he became a secretary in Washington to the Life Saving Service, which eventually became the Coast Guard. It was while he was there that he began to fulfil his ambition to become an inventor, rather as Baird had in the Clyde power station. Jenkins, however, had a more promising idea than the borax-treated undersock. He was a pioneer of the cinema industry, and although his name is not widely known today (he has no entry in the *American National Biography*) he had more than his hour of fame. Like many young men interested in the latest technology, Jenkins had taken up photography. The kinetoscope, a forerunner of the cinema, had become a fairground novelty in the 1890s though it was not a great commercial success. Devised in Thomas Edison's new West Orange laboratory it was made possible by the creation of rolls of celluloid film by George Eastman and chemist Henry Reichenbach in 1889 and marketed under the invented trade name of Kodak. Only one viewer at a time peering down into the kinetoscope could view the Edison peep shows, which were loops of short film playing over and over again.

Jenkins set his mind to creating a projector that could run rolls of film continuously from beginning to end, with a light source to throw the images on to a screen. He had a crude working model by 1894, which he took home to Richmond, Indiana along with a roll of film he had shot of a girl performing a vaudeville 'butterfly dance'. An audience of family and friends saw the show created by what Jenkins called his 'phantoscope'. They were baffled by it and some reportedly searched behind the screen to see if there was some trickery. As he continued to improve his invention Jenkins found he wanted longer rolls of film than were available but he could not afford them. So he went into partnership with another young man he was introduced to when he attended the Bliss School of Electricity in Washington. Thomas Armat had family money and the two applied for a patent together. They demonstrated the projector at an international fair in 1895 without causing much of a stir: the films were the same as those run in the kinetoscope. Then Jenkins quit his job as a clerk to take up inventing full time. He and Armat soon went their separate ways, both trying to improve the phantoscope, and then fell out when Jenkins tried to get a patent as the sole inventor. The dispute was settled with a payment of $2,500 to Jenkins while the patent was sold on to Edison, who marketed an improved version as the Vitascope. Nevertheless, in 1897 the Franklin Institute awarded Jenkins the Elliott Cresson Medal for his phantoscope and rejected attempts by his former partner to have it withdrawn. He is recognised today as the creator of the prototype movie projector which made the film industry possible.

Jenkins tried his hand at a number of ventures before he put his mind to television. There was the steam-driven motor car, which was not a great success, and a high-speed camera worked with a revolving prism instead of a shutter, which was. He was not as hard up as Baird in his early days, but neither was he wealthy. He created his own small laboratory in Washington and put together a young, enthusiastic team. As early as 1913 he tried to exploit the newly invented wireless technology, as Baird had done. When that proved impossible he worked on the scanning and sending of still photographs by wire and then by wireless.

When he began to tackle the problem of sending moving pictures he made a distinction between images sent by wire, which he called television, and images sent wirelessly which he called radiovision. Jenkins was reliant on radio for transmitting pictures yet he appears to have been profoundly ignorant of the history of wireless. In his 1929 book *Radiomovies, Radiovision, Television* he attributes the discovery of wireless telegraphy to the American dentist Mahlon Loomis who claimed to have used atmospheric electricity to send messages twenty miles from one Blue Ridge mountain to another, long before Guglielmo Marconi's first demonstrations in 1896. History has judged this to be mere fantasy. Jenkins makes no mention of James Clerk Maxwell, Heinrich Hertz, Oliver Lodge or Marconi.

Jenkins had an equally homely way of explaining the mechanism of creating radiovision pictures which, he wrote, was 'based on one of the simplest of the mysteries of our childhood, when mother drew parallel lines across a piece of paper under which she had put a penny, and made pictures of the Indian appear. The explanation, of course, is that the lumps on the penny varied the pressure of the pencil on the paper to build up a picture, though the lines were drawn straight.' So much for the problem of scanning for television.

But then, an inventor like Jenkins had no need of much scientific knowledge. Radio was handed to him and, just like Baird, he was reliant on the lively band of American hams who had effectively invented broadcasting by playing each other music when valve sets came in; and the gramophone was on hand with pre-recorded performances. 'Radio vision is not visionary, or even a very difficult thing to do; speech and music are carried by radio, and sight can be so carried just as easily', Jenkins wrote in 1931. This was true, up to a point, and on 13 June 1925 he was able to demonstrate the sending of 'readily recognisable moving objects' from a naval radio station at Anacostia, near Washington, to a receiver in his laboratory in Washington where it was shown to admirals from the Navy, politicians and representatives of the Department of Commerce and Bureau of Standards. In fact, what they saw was not very different from the animated images of the Victorian nursery zoetrope: just a moving silhouette of a windmill.

The only significant difference between the system Jenkins had evolved and that of Baird was that his instruments were well crafted: no biscuit tin for his Nipkow disks but properly worked metal. But what he was showing was not true television: that is to say, the direct transmission of a studio image rather than a filmed animation. He was beaten to that by Baird, and then by the giant communications company AT&T which demonstrated, again with the Nipkow disk system, the transmission of a seated figure both over phone lines and by radio. Nobody was very impressed and the company decided it was not interested in broadcasting.

Jenkins was convinced that radio vision would soon be an entirely new form of entertainment and on 2 July 1928 he established America's first broadcast channel, W3XK, on two wavebands allotted to him by the Radio Commission. As in the early days of radio, there were at first no television sets that could receive his signals so he published instructions on how to make one of his 'Radiovisors'. By 1931 he was able to report: 'At this writing thousands of amateurs fascinatingly watch the pantomime picture in the receiver sets as dainty little Jans Marie performs tricks with her bouncing ball, Miss Constance hangs up her doll wash in the drying wind and diminutive Jacqueline does athletic dances with her clever partner Master Fremont.' Most of those Americans watching these zoetrope-style silhouette productions would have been young radio amateurs and it is an extraordinary thought that something so primitive could be so exciting simply because it was transmitted by radio. Jenkins got a new, broader waveband which allowed him to broadcast more sophisticated movies and anticipated the imminent arrival of Radiovision, 'when we shall see world events as we sit at our fireside'.

Television stations began to spring up all over North America: there were eighteen by the end of 1928. Jenkins and other broadcasters sold some ready-made receivers, but they were expensive and those sold in kit form were more popular. The technology improved but it was still very crude. The Jenkins Television Corporation was formed and attracted investors so that a second station with a more powerful transmitter could go on the air. There were rivals everywhere, including the brilliant Swedish electrical engineer Ernst Alexanderson who was quietly working on a semi-mechanical television system for the giant General Electric Company. It was the GEC system that screened the first televised play *The Queen's Messenger* in 1928. In the same year the Hungarian Dénes von Mihály, who had been working along the same lines as Baird and Jenkins and for a good deal longer, had moved to Germany and demonstrated his system in Berlin. And there was Baird himself, who was now internationally known and, with the existing technology, the star performer.

But the early television boom in the United States petered out. The authorities controlling the airwaves, the Federal Radio Commission and its successor, would not accept that the quality of television pictures was adequate and it would only issue experimental licences. There was no advertising to bring in revenue. Jenkins and others were correct in the optimistic predictions they made about the future of television. But the technology was on the verge of an improvement in quality that they had failed to envisage. To turn his witticism on himself, it could be said that Jenkins did not stay poor enough, long enough. His Television Corporation went into liquidation in 1932, the patents sold off to De Forest and then the Radio Corporation of America, not because they were any use but just to get them out of the way.

* * *

In London, by contrast, John Logie Baird's semi-mechanical system was still achieving television firsts and enjoying a good deal of enthusiastic press coverage. After a first failed attempt, in 1928 he had managed to transmit live pictures of a woman in London to an enthusiastic audience in New York. And Baird was roundly applauded for transmitting the first televised play in Britain when he persuaded the BBC to give him a slot on the airwaves. The play chosen was Pirandello's *The Man with the Flower in his Mouth*, in which there were only three characters. It was produced jointly by Sydney Moseley and a BBC producer Lance Sieveking. An account of the bizarre setting for this broadcast on 14 July 1930 was given some years later by Sieveking. At the time the camera, which could show only head and shoulder shots, had a 'curiously child-like characteristic – it fixed its attention on the interesting object in front of it and wouldn't let go. If the object, or a person, moved away it followed as if hypnotised. When something else came before it it shuddered, blinked and became quite hysterical before it could focus again.'

The picture was black and white but the three actors had to wear grotesque make-up to emphasise their features. The actor Gladys Young who played a woman in a café, the setting for the play, was daubed with blue and dark green and frightened herself when she looked in the mirror. Baird's laboratory and studio had by then moved to 133 Long Acre, Covent Garden and to gain maximum publicity a viewing theatre was created on the roof. Baird had been experimenting with show television in cinemas and halls with a screen composed of a mass of filament light bulbs which glowed at different intensities, creating a grain like that of a wired photograph. An elongated screen of this sort was constructed on the roof of 133 Long Acre and housed in a tent. The play was transmitted from the studio four floors below via a BBC station at Savoy Hill on the north bank of the Thames.

Sieveking recalled: 'The visitors as they arrived were put in a large open-air goods hoist, without railings, and were hauled up to the roof on chain and derrick. . . . Two-thirds of the way through a frantic message came to say that the great screen was overheating and would melt and that the play must stop. I appealed to Baird. "Tell them to go right ahead and let it melt" he said and grinned at me.' When he told the story again on television years later he said the screen was smouldering at the end of the performance.

The newspapers were impressed, but not ecstatic. 'Let it be admitted at once that plays by television are as yet a subject for men of science not for critics of the finer points of acting,' *The Times* correspondent, who watched the play on a Baird Televisor, wrote the following day.

> The visual transmission is far from perfect; you feel yourself to be prying through a keyhole at some swaying, dazzling exhibition of the first film ever made. . . . What the audience sees in the televisor is an image of about the size of a postcard. The clearness of the image greatly and rapidly varies. At its best it allows slow gesture, such as the moving of The Man's hand to his mouth, to make its effect, and even admits of more striking changes of expression. At its worst, when the apparatus appears to have temporarily taken leave of its senses, the whole world of television leaps into the air and the actor oscillates violently between the room above and the cellars beneath, not deigning to pause in the little rectangle on which our attention is fixed.
>
> Perhaps, you think as you watch this dazzling battle between scientific virtue and natural vice, the fellow would behave more discreetly on a larger screen. In this hope, and because a play on a postcard has its limitations as an amusement, we went unofficially on to the roof. Here – an experiment within an experiment – a larger screen, about 10ft square in area, was being used privately so that it must not at present be criticized; but it is permissible to say that, coming out of the long, low tent at the end of which the apparatus was set up, we did feel . . . that we had been watching a cinema of the future.

Baird certainly believed television would play some part in the future of cinema. Another publicity coup was the staging in midsummer 1930 of a show in which interviews were transmitted from the Long Acre studio to an audience in London's Coliseum near Trafalgar Square. A screen composed of 2,100 filament light bulbs stood on the stage and glowed when the theatre was plunged into darkness. There appeared on the screen a series of people who spoke to the audience and listened to questions put by a compère on the stage who had a telephone line to the studio. One of the celebrities was the boxer Bombardier Billy Wells. It was by all accounts a great success. *The Referee* reported: 'History was made at the Coliseum yesterday, when, for the first time in the world's story, a talking picture was transmitted by television; and seen and heard and loudly applauded by vast audiences in the famous theatre in St Martin's Lane.'

Then, with the co-operation of the BBC, Baird managed to televise the finish of the Derby at Epsom, first in 1931 and more successfully in 1932 when the race was transmitted to an audience in the Metropole cinema in central London fifteen miles away. Baird's friend and publicist Sydney Moseley made the point that the cinema audience saw more of the Derby than most people at the Epsom racecourse.

Baird was leading the world. In his efforts to reduce the heat in his little studio he invented what he called 'Noctovision', or television that could see in the dark with infrared rays. All the time he pressed the BBC to allow him to broadcast a regular television service. But as he achieved one 'first' after another there was a growing unease among his backers and those who were close to him. There was a rival technology, which was less advanced than his own in the early 1930s but threatened to make his whirring, mechanical scanning devices, which he had greatly refined, museum pieces.

In his biography of Baird, Moseley asked some of those who had met him in his Long Acre days what they thought of him. In reply in 1951, the feminist writer Rebecca West recalled the time she and the playwright John Van Druten had gone together to Long Acre. She wrote: 'we were impressed by John Baird as a true genius and very delightful to meet'. After the visit the two of them went window shopping around St James's, wondering if they would ever be wealthy enough to buy a painting or some beautiful china. They thought a man of Baird's calibre *ought* to be wealthy enough but sensed that that would not happen because he was 'so obviously the man who sows the seed and doesn't reap the harvest'. Had Baird backed the wrong technological horse?

* * *

In 1957 a mystery Dr X appeared as a contestant on a popular American television show *I've Got a Secret* sponsored by Winston, the cigarette company. A gangling, gaunt man in a suit settled down uneasily next to the compère of the show and faced the panel of four minor celebrities, who eyed him with curiosity. They were not given the man's name in case one of them had heard of him and his secret would be out. It was a little ritual of the game that the contestant whispered their secret to the compère (who, of course, already knew what it was) while it was revealed on screen to the audience at home and in the studio. When it flashed up there was an intake of breath and a round of applause: 'I invented electronic television.'

The compère indicated there was a bit more to Dr X's secret and the audience read on a screen: 'When I was 14 years old – 1922'. (In fact, he was 16 in 1922.) Dr X was hesitant and did not make it easy for the panel. They figured he must be some kind of scientist as he was not a medical doctor or a dentist. Did he use small animals, like mice, in his research? Oddly, he hesitated. The compère helped him out, chipping in with: 'No, not really.' When the compère revealed that this was Philo T. Farnsworth the panel did their best to give the impression that they knew who he was. He was quizzed a little about his work but he was a dull speaker. After a few minutes he was handed his prize money and a multipack of Winstons, 'America's best tasting cigarette', and disappeared behind the studio curtain, back to obscurity after a brief appearance on the gadget he claimed to have invented.

Farnsworth was not a charlatan: he had made a genuine contribution to the development of the more sophisticated system of television that replaced the semi-mechanical versions pioneered by Baird's generation.

He held patents for which he was paid $1 million, after a long battle in the courts with the giant Radio Corporation of America (RCA). But the notion that he 'invented electronic television' single handed, or that the technology that put him on the screen in 1957 was his creation, is absurd. He knew that well enough: when he was asked on the panel show if anyone else was involved in developing television he said: 'Yes, thousands.' But he had had such a David and Goliath battle with RCA that he was perhaps cornered into claiming publicly more than he knew was his due. It is possible that he is better known now than he was in 1957, for his star has risen since his death in 1971. In 1983 he appeared on a 20-cent US stamp with the caption 'First Television Camera'. In more than one book Farnsworth's life has been romanticised and turned into a folk tale with the title 'The Farm Boy Who Invented Television'.

Farnsworth was born in 1906, the eldest of five sons of a Mormon family who farmed in Utah. At that time there were no power stations supplying electricity to most of rural America, but in 1919 the family moved to a 240-acre ranch in the Snake River Valley near Rigby, Idaho which was equipped with a Delco-Light generator. This had been the brainchild of another inventive American, Charles Kettering, as an offshoot of his work on automatic electrical ignition for automobiles. In 1909 Kettering and a partner founded the Dayton Engineering Laboratories Company and under the brand name Delco manufactured electrical generators driven by small two-stroke engines; these revolutionised life in much of rural America until 1936, when a programme of mains supplies came in. As a boy, Farnsworth became fascinated with the generator, and his interest in electricity was sharpened when he found left behind in the attic of the bungalow in which his family lived a great many magazines which had articles on all the latest gadgetry and ideas. Like other children in rural America he was spellbound, too, by the Sears catalogue or *Wish Book*, which listed many electric toys.

In her book *Distant Vision* his wife Elma, who was intimately involved with his work, gave an account of how Farnsworth came to conceive of his television system:

One day he read an article about sending pictures along with sound by means of radio signals through the air. Keep in mind this was in 1919, and radio was still in swaddling clothes. The method described used the Nipkow spinning disc with a spiral pattern of holes through which the image to be transmitted was scanned. To thirteen-year-old Philo, the method seemed clumsy and inadequate. There had to be a better way. Bit by bit he collected information that eventually led him to discover that mysterious, vitally important particle called the electron, the study of which would define his life.

So young Farnsworth had already sprinted ahead of the field at the age of thirteen: although just a farm boy who spent his day ploughing when he was not at school he knew a Nipkow disk would prove to be a failure. And he knew about electrons, which had first been identified by the physicist J.J. Thomson in 1897. His scientific knowledge appears to have been impressively up to date, and while out ploughing one sunny morning Philo had a flash of inspiration, almost like a religious revelation. This is the account in *Distant Vision*:

As usual, his thoughts turned to how he might train electrons to convert a visual image into an electrical image so it could be sent through the air. He knew this had to be done in a vacuum. He had read of a man named Braun who had made a crude vacuum tube and who had produced light by directing an electrical beam to a surface coated with photosensitive material. He had also read that an electron beam can be manipulated in a magnetic field.

As he turned the horses for another row, he looked back along the even rows he had made in the damp earth. A thought struck him like a bolt out of the blue! The tremendous import of this revelation hit him like a physical blow and came near to unseating him. He could build the image like a page of print and paint the image line after line! With the speed of the electron, this could be done so rapidly the eye would view it as a solid picture! He could hardly contain his excitement. After mulling this idea around in his mind

> all this time and piecing it together one piece at a time, it had fallen together like a puzzle!

Why Philo needed a furrowed field to imagine scanning an image is a puzzle, because the technique of building up a picture line by line was already nearly a century old. The very inventive, but ultimately unfortunate, Scottish clockmaker, Alexander Bain, had actually built a working facsimile machine in the 1840s. And the possibility of using some version of Braun's cathode ray tube had been the subject of a much-discussed article by the eminent Scottish engineer A.A. Campbell Swinton, published in *Nature* magazine in 1908 and expanded on in a lecture of 1911. Dated 12 June, his letter in *Nature* was in response to another, which had cast doubt on the very possibility of television. It began:

> May I point out that though, as stated by Mr Bidwell, it is wildly impracticable to effect even 160,000 synchronised operations per second by ordinary mechanical means, this part of the problem of obtaining distant electric vision can probably be solved by the employment of two beams of kathode rays (one at the transmitting and one at the receiving station) synchronously deflected by the varying fields of two electromagnets placed at right angles to one another and energised by two alternating electric currents of widely different frequencies, so that the moving extremities of the two beams are caused to sweep synchronously over the whole of the required surfaces within one-tenth of a second necessary to take advantage of visual persistence.

He went on to make the point that there would have to be more discoveries before his idea could be made to work but that it seemed theoretically possible.

A year after the letter was published, an account of the experiments of a German professor, Max Dieckmann, who used a cathode ray tube to display images, was published in *Scientific American*. And in 1915, when

Farnsworth was nine years old, the great populariser of radio in America, Hugo Gernsback, published an article in the magazine *Electrical Experimenter* on what he called the 'Campbell Swinton electronic scanning system'.

The notion that 'electronic television' was the brainchild of a Mormon farm boy is evidently preposterous. What is extraordinary, however, is the fact that Farnsworth set about the almost impossible task of making it a reality though his scientific education was limited and he had no money of his own at the outset. He had not completed studies at Brigham Young University when his father died suddenly. Philo was eighteen, and as the family's breadwinner he decided to join the Navy where, to his annoyance, he was given the nickname Fido. He left after two months when he learned that any inventions he might make while he was in service would belong to the Navy. Back home he had a variety of jobs, then returned to university. He also met his future wife Elma Gardner, a school friend of his sister. In 1925 he and Elma's brother, Cliff, moved to Salt Lake City in search of better-paid work and Phil (as he now called himself) got a job as a radio repairman.

If there was any divine hand at work guiding Farnsworth towards his destiny, as Elma suggested, it showed itself in Salt Lake City. Two men from San Francisco who made a living as fund-raisers, George Everson and Leslie Gorrell, were in town to find money for a local community chest. They recruited Phil and Cliff, and became friendly with them. When they heard of Farnsworth's burning ambition to create electronic television they explored the possibility of providing him with the funds to start a laboratory. Everson had some savings, but not sufficient to finance a research programme. But he and Gorrell were fund-raisers, after all, and they offered to see if they could raise money for Farnsworth's project. Everson suggested that Farnsworth move to Los Angeles where they would have a greater chance of raising money and where technical expertise might be more readily available. Farnsworth agreed, on condition he could bring along his girlfriend Elma, always known as Pem. So they married before heading west. In Hollywood they rented an apartment and in June 1926 set about putting together a home-made laboratory. Everson and Gorrell gave them $150 a month to live on and to buy some materials, while they went in search of more substantial funding.

There was one piece of equipment they could not make themselves: the vacuum tube. Farnsworth found a scientific glassblower who did his best to produce a tube to the inventor's specifications. The very first camera was assembled at the glassblower's workshop and carried back to the flat. Farnsworth called it his 'image dissector' as its function was to scan a picture and convert light into electricity. With his backers looking on, he switched on the power. It blew up.

This might have been the end for the Farnsworth all-electric television system had it not been for the dedication of Everson and Gorrell. They had found financial backers among the people they knew in California – despite the fact that Farnsworth was only nineteen years old and had absolutely nothing to show his backers but boundless enthusiasm and almost crazy ambition. A syndicate based on the Crocker National Bank put up $25,000 and provided laboratory space in San Francisco in return for 60 per cent of profits, the remaining 40 per cent to be divided equally between Everson, Gorrell and Farnsworth. In September 1926 they moved into their new premises at 202 Green Street just by Telegraph Hill. They were joined by Cliff Gardner, now Farnsworth's brother-in-law.

It is perhaps not as surprising as it might first appear that hard-nosed businessmen were prepared to gamble a significant sum of money on such an unlikely project. In 1926 Jenkins and Baird were beginning to receive a great deal of publicity. Seeing by wireless was a new wonder of the age. The fact that pictures were far too crude to be commercial was not a great concern as by that time everyone was used to new inventions improving very rapidly. Mechanical television looked as though it was about to break through. And if that did happen, then the superior picture quality Farnsworth sought could be worth a fortune. Invention and innovation have invariably progressed in this way, with the first, crude prototypes serving the vital function of showing that something can be done. When these embryonic forms of an invention become obsolete their creators are liable to be dismissed as misguided or backward. But it is their pioneer work that generates optimism and draws out the backing for the more advanced technologies which replace them.

However, when Farnsworth began his labours, the industry view was still that some version of mechanical scanning of images was the most promising way forward for the transmitter. It was not that people were unable to see that a cathode ray tube camera would be superior. That was not in question. Very little work was being done on it because the problem of discovering how to manipulate electrons in a vacuum tube involved experimentation with a much less accessible technology than the Nipkow disk or any of the other mechanical scanners. With the mechanical scanners you could pretty much see what was going on with the naked eye. How electrons were behaving in a sealed glass tube was not apparent and involved a highly sophisticated understanding of physics.

The key discovery that electrically charged particles a thousand times smaller than atoms would travel through a vacuum had been made in the nineteenth century. It had been shown, too, that when these particles, or electrons as they became known, hit a photoelectric surface they could produce an image. Farnsworth's all-electric television system would have to manipulate this laboratory equipment in some way so that the cathode ray tube performed the same function as the Nipkow disk and the selenium cell. He was not the first to attempt this but he had a chance to be the first to make it work. The biggest problem was with the camera.

As Everson had anticipated, California provided Farnsworth with some much-needed expertise as well as financial backing. Bill Cummings, in charge of glass blowing for the University of California in Berkeley, who had made them their first tubes, taught Cliff Gardner the art. In time he became very skilful. Meanwhile, Everson and Pem worked with Farnsworth making magnetic coils and experimenting with the photosensitive materials. At the outset, Farnsworth was wildly optimistic about what he could achieve in a short space of time. In 1927 he took out his first patent, but he did not have anything like a working system. His first public showing of all-electric television in 1928 was not very impressive. Not long after that his laboratory was destroyed in a fire. If they were to keep going, money had to be raised on the open market so Farnsworth Television was incorporated and shares sold.

While Farnsworth struggled, semi-mechanical television was enjoying one triumph after another. The quality of the image was improving, with the use of a 'flying spot' scanner in place of the Nipkow disk. Broadcasting as such had not begun, but Baird and his backers were putting pressure on the British Broadcasting Corporation to give them airtime. Those who had invested in Farnsworth were understandably anxious that their man was not so much ahead of the game as left at the starting blocks. It was at this critical point in the development of television that Farnsworth was visited by a man he had not met before, but one he knew of by his writing and reputation. They were of the same mind as far as all-electric television was concerned and Farnsworth had nothing but admiration for him.

* * *

When Philo Farnsworth was just one year old a patent for a television system which used a cathode ray tube as a receiver was filed in Germany and Britain by a Russian scientist, Boris Rosing. That was in 1907. In 1911 Rosing applied for another patent for a more sophisticated system which had been developed with the help of a young assistant called Vladimir Zworykin who had enrolled as a student at the St Petersburg Institute of Technology in 1906. Zworykin, the scion of a wealthy family, was born in Murom, western Russia, in 1889. He had an idyllic childhood, horse riding and exploring along the Oka River, and studied at the local gymnasium. Rosing recognised Zworykin's aptitude for science and while he was still a student involved him in his research into cathode ray tubes. When Zworykin graduated in 1912 his father wanted him to join the family business: he could either do that or he must go abroad. Zworykin chose to go to Paris, where he studied X-ray technology at the Collège de France. From there he went on to Berlin. When war broke out in 1914 he was in enemy territory and was lucky to make it back to his home country. That was the end of Zworykin's idyllic childhood and youth: for the next decade he was unable to do any scientific work or to further his studies.

Zworykin got back to Russia, escaping through Denmark and Finland, and joined the Army, where his expertise in radio communications was recognised. He then moved to St Petersburg (briefly renamed Petrograd during the First World War to rid it of the Germanic 'burg'), where he was commissioned as an officer and worked in the Russian Marconi factory. He planned to return to his research on television there when the war was over. Meanwhile he continued to work on radio communications for the government, and got married. As the son of a wealthy businessman he was in danger when the Russian Revolution began and he and his wife Tatiana went into hiding in Germany before the war was over. Tatiana stayed in Berlin while he went back to Russia to see if he could again work for the Marconi company. But everything had changed: the company was more or less idle and his family home had been requisitioned. Bitter infighting between the revolutionaries was tearing the country apart. So, in 1919, Zworykin took a ship to America. He spoke little English and could get only a menial job so he returned to Russia. It was not long before he turned tail again and was back in America, determined to learn English. Tatiana joined him in New York.

Zworykin got a job with the Westinghouse Electric Company, leaving after a short while because he was not happy with the pay. However, he was soon rehired on much better terms and given his own lab; he had finally found his feet again. Westinghouse had no interest in developing television but Zworykin had time to experiment on his own and began work on a camera tube. In 1923 he applied for a patent for a working television system. The following year he became an American citizen and in 1926 was awarded a PhD in physics by the University of Pittsburgh. He continued to work for Westinghouse on a variety of projects and as he gained a reputation for his inventiveness with cine cameras and facsimile machines he finally got the chance to do what he really wanted to do: his ambition was exactly the same as Farnsworth's.

In time he came to the attention of David Sarnoff, then vice president of the all-powerful Radio Corporation of America (RCA). Radio had rapidly become big business and, like others in the industry, Sarnoff kept an eye on developments in television. He recognised that existing technology was not offering anything potentially commercial and he arranged for Zworykin to visit Europe, as he had done himself, to see if there were developments that might make all-electronic television possible. Baird, still wedded to mechanical scanning, was of no interest to Sarnoff or Zworykin. However the Laboratoire des Etablissements run by Edouard Belin in Paris most certainly was. Belin had a team working on cathode ray tubes and they had made discoveries about the behaviour of electrons under different conditions that were a revelation for Zworykin. A new kind of cathode tube pumped out to a high level of vacuum made electrostatic manipulation of electrons possible, and he was able to bring back samples to work on in America. He also found a talented fellow Russian, Gregory Ogloblinsky, who was willing to join him at Westinghouse. That visit to Paris in 1928 is regarded by historians of television as the crucial breakthrough as it enabled Zworykin to go on to develop a cathode ray camera tube, which brought in the first phase of high definition television.

Sarnoff had also heard of Philo Farnsworth and his attempts to create all-electric television. So in 1930 he suggested that Zworykin see him as well. He was warmly greeted by Farnsworth and his team, and spent four days with them in the Green Street laboratory. He was especially interested in the image dissector, Farnsworth's camera tube, and admired the way the glass had been moulded so that there was a flat rather than a rounded end. In this otherwise cut-throat industry it is surprising that Farnsworth was so trusting. In *Distant Vision* Pem says he was misled: he thought Zworykin was working for Westinghouse and that there could be a lucrative licensing deal with them, whereas he had been sent by Sarnoff at the RCA. Certainly Farnsworth's backers were alarmed when they learned that Zworykin had not only taken an image dissector but had got Cliff Gardner to show him how to make one.

It is impossible to say how much Zworykin borrowed from Farnsworth when developing his own television camera, which he called an iconoscope. The team of scientists working with Zworykin certainly made their own versions of the image dissector and attempted to improve it. And David Sarnoff visited the Green Street laboratory after Zworykin had been there and offered to buy out all Farnsworth's patents for $10,000. As it happened, Farnsworth was not there when Sarnoff turned up but when he heard of the offer he was not impressed. He believed he would make more than the small fortune Sarnoff was offering once the new industry was clamouring for his all-electric system.

Farnsworth had a small team of technicians and scientists working with him. But once the Radio Corporation of America decided to get into the business of developing television it was going to be difficult for him to compete. Rebuffed by Farnsworth, David Sarnoff decided he did not need him or his image dissector. Zworykin had one anyway and the team he was putting together to improve on it made many versions. Designing a cathode ray camera tube proved to be extremely difficult. The theory about the behaviour of electrons in a vacuum was really not much help. Whereas Baird's semi-mechanical system with its flying spot scanner was about as good as it would ever get, the all-electric approach to television was still experimental.

It is really impossible to give a coherent description of how Zworykin and his team produced the camera tube that was to make all-electronic television – the mechanical scanner replaced by a vacuum tube – commercial. Accounts of the process suggest that there was a great deal of trial and error and not much theory. Many configurations of tubes were tried and many different versions of the photosensitive mosaics which would form the picture. One critical breakthrough happened completely by accident. A great deal of experimentation went on in search of the most effective photoelectric 'mosaic' in the tube, which involved heating a silver surface on a mica plate. A member of the research team left one of these silver mosaics in the oven longer than intended and found that it was vastly improved. This led to the team's discovery of the principle of 'charged storage' which gave the tube much greater brightness.

Finally Zworykin was able to unveil what he called his iconoscope, which rivalled Farnsworth's image dissector and, ultimately, improved upon it. Once the iconoscope's workings were revealed there were a number of claims for precedence from researchers in Europe. But it soon became clear that the big battle was going to be between the semi-mechanical system that Baird was promoting on an international scale and the all-electric system developed by RCA.

This was played out most dramatically back in London. Whereas the Americans had rejected semi-mechanical television, the newly founded British Broadcasting Corporation, which had a monopoly, decided to conduct an experiment in which Baird would be able to compete with an all-electronic system.

* * *

While Zworykin and his researchers were puzzling over the workings of the iconoscope, John Logie Baird had finally found a wealthy financial backer. In January 1932 the financier and film producer Isidore Ostrer acquired control of the Baird Television Company. Born in the East End of London in 1889 to Jewish immigrants who had fled pogroms in the Ukraine, Ostrer had begun work as a humble stockbroker's clerk in the City, and later set up a merchant bank with two of his brothers.

They moved into textiles and then, in 1927, just as the 'talkies' came in, created the once dominant chain of cinemas: the Gaumont–British Picture Corporation. Ostrer had 350 cinemas within a few years and tried his hand at film production, but this was not a great commercial success. Investment in television seemed a natural enough progression for Ostrer, and Baird was struggling, cutting his staff back to about twenty in 1931. After five years of investment the company had leased an area of the Crystal Palace at Sydenham in south London in 1933 to create a studio and experimental television station. By 1936 it had 382 employees.

Baird had been transmitting experimental programmes using the BBC transmitter when there were no radio programmes on air: late at night and in the morning. The BBC indicated that these would come to an end and that a judgement would be made about whether or not it was interested in pursuing television. In order to explore the technical issues and future prospects of this new medium, the Government set up a committee of enquiry in 1934 chaired by Lord Selsdon. A rival to Baird Television had demonstrated its system to the BBC with favourable results. This was the company that became known as Marconi-EMI, which had a vital link to the RCA in America and Zworykin's latest technology. Electrical and Musical Industries (EMI) had been formed in 1931 by a merger of the Columbia Graphophone [*sic*] Company and the Gramophone Company with its record label HMV.

An employee of the Columbia Company was Isaac Shoenberg, a brilliant engineer and mathematician. Like Zworykin, Shoenberg was Russian, born in Pinsk in 1880 into a Jewish family involved in forestry. He studied mathematics and electrical engineering in Kiev and found work in the new wireless industry, joining the Marconi Company in 1911. He married young, at just twenty-three, and already had four children when in 1914 he moved the family to London, which he had visited as a Marconi chief engineer. When war broke out he was turned down for military service and joined the British Marconi Company where he became head of the patents department. Shoenberg loved music and took the offer of a job with the Columbia Graphophone Company. When this company merged with the Gramophone Company, which was developing television, Shoenberg was made head of research at the new EMI and drew together a richly talented team of scientists and technicians. Of the 114 staff, thirty-two had university degrees; nine of these were doctorates.

EMI inherited business links with the RCA and it was not long before Shoenberg was adopting and adapting Zworykin's iconoscope and developing his own version of all-electronic television. The EMI camera tube became known as the Emitron. The Marconi-EMI system was soon threatening to make Baird's system obsolete. Picture quality was defined in part by the number of lines in the scanning of the pictures. In the mechanical devices these were limited because the rotating Nipkow disk had a speed limit. For a long time Baird had worked with just 30 lines. The Emitron camera held out the possibility of 405 lines.

It was too late for Baird's technicians to try to compete with EMI. They simply did not have the expertise. In desperation they contacted Phil Farnsworth, who, in 1934, immediately packed up his equipment and shipped it to England, taking with him two engineers. His liner did not dock at Southampton and he had to take a pilot boat with his equipment lowered on to the deck. Farnsworth was wined and dined by the Baird Company while the equipment was unpacked at the Crystal Palace studios. After a bit of horse trading, Farnsworth managed to get $50,000 for his image dissector. However, it did not save the day for Baird as it was not as finished a piece of equipment as his company had been led to believe.

The Selsdon Committee decided to hand over the task of introducing the first public television service to the BBC and recommended that both the Baird and Marconi-EMI systems should run the station on alternate weeks. Baird won the toss of Lord Selsdon's coin to take the first week. For its new television centre the BBC had chosen to convert part of a Victorian pile, Alexandra Palace, which stands on a hill with a fine view over London. The first programmes were aired on 2 November 1936 and were generally favourably received, the *Radio Times* offering readers a new column by 'The Scanner'. Whether or not these broadcasts are to be considered the first showing of public television in the world has long been disputed.

The year before, the Nazi regime had introduced a television service using a semi-mechanical system borrowed essentially from Baird. Naturally the station in a tiny room with its scanning disk was named after Paul Nipkow, the German 'inspirer of television'. Hitler insisted on closing down all foreign companies but the Germans hung on to the television equipment developed in Britain and America. Television was strictly for propaganda purposes, with presenters greeting viewers with the Nazi salute and a 'Heil Hitler'. Introducing the service Eugen Hadamovsky said it was there to fulfil the 'sacred mission of imprinting the image of the Führer on every German heart, never to be erased'. There were no domestic television sets, only 'television parlours'. The first broadcasts in 1935 did not attract much applause: the pictures were no better than those Baird had broadcast five years earlier. It was not until coverage of the 1936 Olympics that there was anything like an enthusiastic audience in the television parlours. Exceptionally sharp pictures were achieved by film cameras mounted on vans: engineers rapidly developed the footage, which was then scanned and transmitted to viewers.

It became clear in Britain that the Baird television system was not performing as well as the Marconi-EMI all-electronic system with the Emitron camera. Then, on 30 November, less than a month after the first programmes went on air, the Crystal Palace, which housed Baird's equipment and studios as well as Farnsworth's equipment, was burned to the ground in a spectacular blaze which lit up a large part of southern England. The following year the BBC dropped Baird altogether. It was a great blow for Baird felt he was the one who had battled with the Corporation and had persuaded them to give television a chance. However, the Board of Baird Television took a more optimistic view: they could make more money making television sets than going into broadcasting. Then the war came and the BBC television service was closed down, the funds and expertise deployed on more important work.

Baird might have backed the wrong technology but amongst those who strove to prove that seeing by electricity *was* possible he was the most influential in the very early days of the technology. His achievement can be compared with that of the Wright Brothers, whose early Flyers were rapidly superseded by more sophisticated and efficient aircraft. Yet John Logie Baird died in 1948 a disappointed man. His name had been almost completely erased by the BBC and it was not until a long time after his death that his contribution to the long history of television was properly acknowledged. He outlived Paul Nipkow by only eight years, the designer of the scanning disk dying in Berlin in 1940.

Philo Farnsworth, like Baird, was written out of the history of television for many years and he too felt bitterness about his treatment. He gave up television and tried his hand at nuclear fusion without success. He died in 1971. Tragically, the Russian physicist Boris Rosing, who had first fired Zworykin's interest in television and held two of the earliest patents, was branded a 'counter revolutionary' in 1931 and exiled to a remote district where he died the following year of a brain haemorrhage. Zworykin, on the other hand, lived to the age of ninety-two. But he was not overjoyed by the realisation of the dream of being able to 'see with electricity' which he had helped to bring about. Asked what he thought of television he said simply: 'Awful'.

Television is one of those inventions that has never been welcomed or admired universally: the sophistication of the technology is no guarantee of the quality of the programmes. In contrast, the now ubiquitous bar code, which has insinuated itself into our lives without fanfare or engendering any sense of wonder, is surely an invention which has been accepted wholeheartedly. The concept was simple enough but the technology that made it practicable was based in part on a revolution in our understanding of physics.

CHAPTER 3

WRITTEN IN THE SAND

Joe Woodland said himself it sounded like a fairy tale: he had got the inspiration for what became the bar code while sitting on Miami Beach. He drew it with his fingers in the sand. What he was after was a code of some sort that could be printed on groceries and scanned so that supermarket checkout queues would move more quickly and stocktaking would be simplified. That such a technology was needed was not his idea: it came from a distraught supermarket manager who had pleaded with a dean at Dextrel Technology Institute in Philadelphia to come up with some way of getting shoppers through his store more quickly. The hold-ups and the regular stocktaking were costing him his profits. The dean shrugged him off, but a junior postgraduate, Bernard 'Bob' Silver, overheard and was intrigued. He mentioned it to Woodland, who had graduated from Dextrel in 1947. Woodland was already an inventor and he decided to take on the challenge.

So confident was he that he would come up with a solution to the supermarket dilemma that Woodland left graduate school in the winter of 1948 to live in an apartment owned by his grandfather in Miami Beach. He had cashed in some stock market profits to tide him over. It was in January 1949 that Woodland had his epiphany, though the brilliance of its simplicity and its far-reaching consequences for modern existence were not recognised until many years later. It was only when the bar code had become a commercial reality that Woodland was asked to tell his story, as he did to the *Smithsonian Magazine* in 1999, the fiftieth anniversary of his inspiration.

It was Morse Code that gave him the idea. Woodland had learned it when he was in the Boy Scouts. As he was sitting in a beach chair and pondering the checkout dilemma, Morse came into his head:

> I remember I was thinking about dots and dashes when I poked my four fingers into the sand and, for whatever reason – I didn't know – I pulled my hand toward me and I had four lines. I said 'Golly! Now I have four lines and they could be wide lines and narrow lines, instead of dots and dashes. Now I have a better chance of finding the doggone thing.' Then, only seconds later, I took my four fingers – they were still in the sand – and I swept them round into a circle.

Back in Philadelphia, Woodland and Bob Silver decided to see if they could get a working system going with the technology to hand. They first filed a patent in 1949, which was finally granted in 1952. Although the patent illustrates the basic concept there is only a smattering of anecdotal evidence about what Woodland and Silver actually built. There was no modern scanner technology available to them, and no microcomputer to decipher information. However, Woodland had previously devised a way of playing up to fifteen tunes in sequence in elevators. He had borrowed a system for imprinting sound on film devised by Lee de Forest, which involved turning sound into light, which was then turned back into sound with a photoelectric cell. Woodland reasoned that the same technology could be used to read bar codes, provided a strong enough light were shone on them. A crude prototype in Woodland's own home used a powerful 500 incandescent bulb. An oscilloscope was used to 'read' the code. The whole thing was the size of a desk. Allegedly, it worked, up to a point. But an objective evaluation judged it to be twenty years ahead of its time.

Woodland believed in it, however, and joined the huge business machine company IBM in the hope that they might develop it. But IBM was not interested. At that time, in the 1950s, none of the big technology companies, in fact, showed the slightest interest in the concept of the bar code. It was another classic case in the history of invention in which an outsider sees the future more clearly than those engrossed in existing and lucrative technology. In another sense, the invention of the bar code, and its astonishing proliferation, is untypical in that it provides a rare instance of necessity eventually being the mother of invention. Nobody was looking for a television before it appeared in the world, any more than they were desperate to fly, or to wander around with a telephone. These inventions *created* a demand that only a few visionaries imagined was there. However, the bar code came about to solve problems faced by industry and the retail trade as the range of products was becoming unmanageable. The first truly practical application of this invention was the result of an extraordinary array of talents, from Nobel-Prize-winning physicists to supermarket managers. The theories of Albert Einstein had a hand in it; although, as he died in 1955, he did not live to see the little barred footprint on his groceries. It was in July 1974, after much experimentation and negotiation, that the true, Universal Product Code was first scanned in a supermarket: a ten-pack of Wrigley's fruit gum which is now an exhibit in the Museum of American History in the Smithsonian.

The bar code, more than any other invention, somehow joins the sublime and the ridiculous: quantum mechanics, the checkout and chewing gum. Woodland and Silver had the right idea but they lacked the minicomputer and, critically, a very bright light with which to 'read' the black and white bar code.

* * *

On 17 July 1960, at a press conference held in the Delmonico Hotel in New York, one of the most sensational announcements in the history of science was made by the Hughes Aircraft Company of Culver City,

California. One of their research scientists, Dr Theodore Maiman, had made an 'atomic radio light brighter than the center of the sun'. Maiman produced for the newsmen his 'laser', an acronym for *light amplification by stimulated emission of radiation*. What Maiman held in his hands did not look that impressive. In fact, when he first saw Maiman's laser the night before the press conference, the head of public relations at Hughes, Carl Byoir, declared they were in big trouble: 'It looks like something a plumber made.' According to Maiman in his book *The Laser Odyssey*, the science writer for *Time Life* slammed the press kit on the table and shouted: 'What kind of a hoax is Hughes trying to pull here!'

Most of the reporters, however, were eager to learn what the laser was for, and what it could do. It was like science fiction. Maiman said the laser beam was so concentrated, so 'coherent', that if it were beamed from Los Angeles to San Francisco it would spread only 100 feet. The tiny beam was hot and sharp enough to cut through materials. Could it be used as a weapon? That was not the intention, Maiman assured reporters. Nevertheless, the *Los Angeles Herald* headlined its story: 'LA Man Discovers Science Fiction Death Ray.' This became a popular theme in the newspapers.

Maiman had won the race to build the very first laser, beating fierce competition from around the world. He, like all the other scientists exploring the possibility of finding a practical way of amplifying a beam of light, was pursuing a line of enquiry first proposed in 1917 by Albert Einstein in his theoretical account of the behaviour of particles under certain conditions. The paper was entitled 'On the Quantum Theory of Radiation' ('Zur Quantentheorie der Strahlung') and predicted the stimulated emission of electromagnetic radiation. Over the next half-century scientists confirmed in laboratory experiments that Einstein's theory was correct.

After a widespread international investigation of the possibilities of 'stimulated emission' the first breakthrough came in the United States in the 1950s with the creation of the first maser, again an acronym, meaning *microwave amplification by stimulated emission of radiation*. This was the audio-only version, which has had important value for scientific investigation. Its creation raised hopes for the possibility of producing even shorter microwaves, which would give a highly concentrated beam of light. The principle was known to a large number of scientists: a beam of light is shone on a medium of some kind in which the electrical charges become 'excited' so that they produce what is called a *coherent* beam as opposed to the *incoherent* beam of a conventional light. One of the substances used in research was man-made ruby. After many experiments with it, the scientific wisdom in the late 1950s was that it would not work. Theodore Maiman, however, had faith in it. In *The Laser Odyssey* he describes the day it actually worked for the first time:

> It was the afternoon of May 16 1960; it was time to confirm or deny all of the fears of why the 'ruby can't work'; Or, why 'lasers can't be made to work'. No more calculations, no more diversionary experiments, this was the moment of truth!
>
> The laser head was mounted on a workbench. The flashlamp was connected to the power supply. The trigger electrode was connected to the spark coil [the mechanism that initiates the flash from the strobe lamp]. The light output from the coupling hole in the end of the ruby was directed through a Bausch and Lomb monochromotor to a photomultiplier tube [a very sensitive form of photoelectric cell]. . . .
>
> We took a test shot so that we could adjust the monitoring equipment. We turned up the power supply to about 500 volts. . . . We fired the flashbulbs. . . . We progressively increased the supply voltage, each time monitoring and recording the light output trace. As we did so, the peak output increased proportionately to the energy input and decay time remained the same. So far, so good. *But when we got past 950 volts on the power supply everything changed! The output trace started to shoot up in peak intensity and initial decay time rapidly decreased. Voilà! This was it. The laser was born.*

It reads like science fiction and is unintelligible to the lay reader. But it is possible to imagine the extreme excitement that Maiman and his associate Irnee D'Haenens experienced when they produced that first, fickle beam. They did not know then what it might be used for, but they imagined it would have many applications in science and communications, in industry for cutting and welding, and in medicine for delicate surgery. But, as Maiman wrote, 'I did not foresee the supermarket check-out scanner or the printer.'

As so often with new inventions, once the laser was created it proliferated rapidly and there were soon models available for those experimenting in many different fields. One of these would be the bar code: the laser was the intense light Woodland and Silver had needed. Their 1952 patent, which had languished, was bought in 1962 by the company Philco for an undisclosed sum. Two years later, when he was just thirty-eight years old, Bob Silver died. Woodland continued to work on the bar code concept at IBM and was later to see his original concept developed by a rival.

The first of the technological breakthroughs which created the familiar bar code became available from the early 1960s. But technology alone could not bring the bar code to life. All inventions require a degree of industrial co-operation and entrepreneurial enthusiasm to become commercially available. The bar code is exceptional, however, among modern innovations: it has to be approved of and agreed to by a committee or it is of no value at all. And that committee will comprise representatives of companies which are otherwise fierce competitors. It is a most extraordinary case of innovation by co-operation, and it began not in the supermarket but along the railroad tracks of North America.

* * *

In 1957 David J. Collins, who had newly qualified as an engineer and had been accepted to study at Massachusetts Institute of Technology on a management programme, found he had time to kill in the summer. He fulfilled several weeks of army reserve duties before he accepted the offer of a place on a training programme with the Pennsylvania Railroad Company. When he was growing up he had heard a great deal about railroads because their troubles and inefficiencies were the topic of conversation around the dinner table. His father had a company that sold equipment to the railways. One of the biggest problems that nobody had been able to solve was how to keep track of goods wagons (railroad cars in America), which were hauled all over the country. They were always disappearing. In the late nineteenth century a patent was taken out for a mechanical system that would automatically log the whereabouts of the cars, but it did not work.

When he graduated from MIT, a stimulating course which involved an exchange spell in England – where he became a temporary lieutenant commander in the Royal Navy with responsibility for torpedo launching refinements – Collins had no desire to work on the railways. Instead he took a job with the well-established electronics company Sylvania. He joined a research team of around 150, more than half of whom were well qualified postgraduates. Sylvania was on the lookout for projects that might make use of a large computer that had been acquired for communications research for the military. Collins suggested they develop a system to track railcars automatically which would involve some kind of coding system, a scanner and a brain to store and log the information fed in. It turned out that the Sylvania computer was not suitable, but a basic bar code reading technology emerged. According to Collins it owed nothing to Woodland and Silver's system as he did not know about it until Sylvania started to apply for patents in the same area. The checkout system was not, anyway, applicable to the railcar problem.

Development of what became known as the KarTrak technology began in 1959. Discussions with railroad companies provided a brief: any labels on cars would have to be inexpensive (about $1); scanning of cars should be possible at speeds up to 60 mph; labels should be made robust enough to last for seven years; the scan height on the trackside should be nine feet for most traffic; lastly, the system should be operational in remote areas, and 'rifle resistant' in hunting areas. Work began before lasers were available commercially and the KarTrak system was based on an older technology.

Collins and his team at Sylvania came up with a rugged, practical bar code reading system. The labels were coloured strips of reflecting material which represented coded information about the ownership of the car and made it possible to track its movements. The scanner was a brilliant white xenon light of the type used in film projectors, which picked up the reflection from the reflective strips. This information was then fed into a computer. It was a gigantic system compared with later bar code technologies but the principle was similar.

The engineers and scientists at Sylvania were aware early on of the possibilities presented by the laser but it was not suited to the KarTrak system, which had to operate out of doors in all weathers with a scanner set a considerable distance from the coloured bar code. KarTrak was first tested in 1961 on the Boston & Maine Railroad on a 'captive' gravel train where the cars remained in a confined area. The system worked well enough but it could not be extended to the open rail network until a sufficient number of the companies in the Association of American Railroads agreed to adopt it. To keep production going and get the KarTrak brand well known, Sylvania initially sold only to operators of 'captive' networks: the market was worldwide and a total of 50,000 cars were labelled between 1963 and 1966. But the major rail networks in North America were not convinced.

In frustration with the prevarication, Sylvania established a Rail Data Corporation with a full-page advertisement in the *Wall Street Journal* promising the railroad companies a saving of $3 billion. A plan was sent to the presidents of the top twenty railroad companies, who together owned three-quarters of all railcars. Once half the companies had signed up to KarTrak, Sylvania proposed to label all railcars and provide scanners anywhere. Subscription would be $10 a car and scanner revenue a few cents per recorded 'incident'.

Still unconvinced, the Association of American Railroads decided to invite bids from a number of companies and these were tested in 1966 and 1967. KarTrak was the winner and became the mandatory car identification system in North America, leading to the labelling of 1.5 million cars and the installation of around a thousand scanners.

1. Modern inventions are often anticipated long before the technology evolves to turn fantasy into reality. This American trade card advertising a penknife dates from the 1890s when the wonders of the telephone suggested that one day it would be possible to send pictures down a wire as well. Television arrived sooner than anticipated: not a century later but less than thirty years.

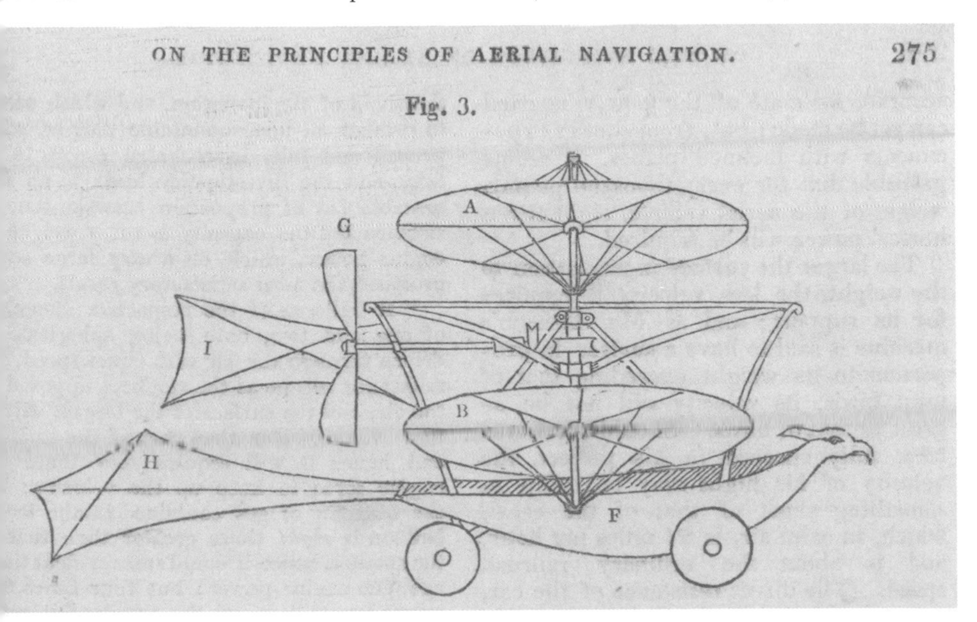

2. One of the more ambitious designs in 1843 by the pioneer of aeronautics the Yorkshire baronet Sir George Cayley. As the Wright brothers acknowledged Cayley had worked out the principles of winged flight a century before they were able to take to the air in a powered flyer.

3. The only known photograph of Si
George Cayley taken in 1844 when
he was 71. In the era of the hot air
balloon Cayley risked ridicule for his
conviction that heavier-than-air fligh
was possible.

4. A dramatic photograph taken in 1894 of the German Otto Lilienthal preparing to launch himsel on one of his many experimental flights. It was the tragic death of this fearless pioneer aviator that inspired the Wright brothers to continue with the quest to fly. They regarded Lilienthal as 'a great man'.

5. Lilienthal taking off in 1893 from a structure he built for his test flights. The following year he created his famous *Fliegerberg* or Flyer's Hill where he would entertain crowds and promote the sport of gliding. He attempted to steer and keep his balance by swinging his legs and his body from side to side but he never got real stability in the air.

6. Unique in the history of invention, a eureka moment captured on film. Orville Wright takes to the air on the 17 December 1903 at Kitty Hawk on the remote coast of North Carolina. Orville's older brother Wilbur is alongside the powered flyer which was airborne for only a few seconds. John T. Daniels, a 'surfman' from the nearby life-saving station took the picture.

7. The Wright brothers could not have become the first to fly without the help of the people of Kitty Hawk. It was a letter from Bill Tate, who ran the Post Office with his wife Addie, that convinced Wilbur Wright Kitty Hawk was the ideal testing ground. When Wilbur first arrived the Tates put him up and fed him. Addie is seated with one of their daughters who had dresses made from the wing cloth of an abandoned glider.

8. The Wright brothers on the porch of their home in Dayton, Ohio in 1909 when they were at the height of their fame in both America and Europe. Wilbur (on the left) was four years older than Orville but their father described them as 'like twins'. They were always dapper dressers and charming in company but neither married and there is no record of any relationship with women. Three years after this photograph was taken Wilbur died at the age of 45.

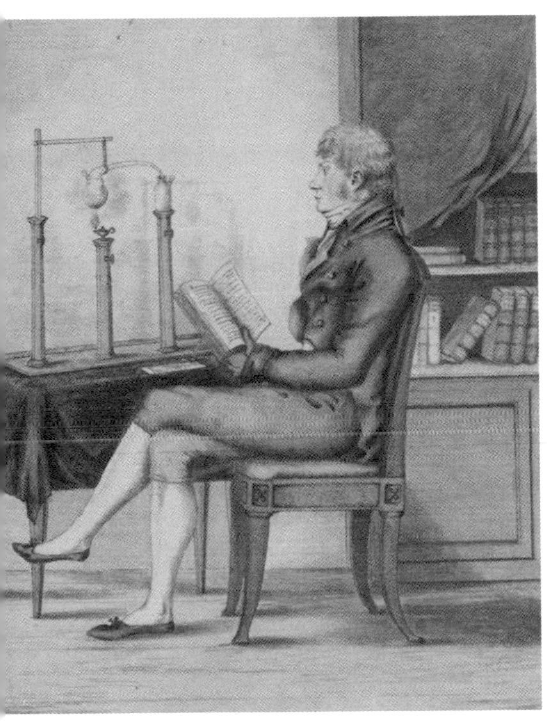

9. A rare portrait of the Swedish chemist Jons Jacob Berzelius who, in 1817, identified a new element which he called *selenium*, meaning 'of the moon'. One of his trusted pieces of equipment, used for analysing rock samples, was a blow pipe, which was used to produce intense heat from a candle flame. Long after his death it was found that selenium was sensitive to light and it provided the first breakthrough in the creation of television.

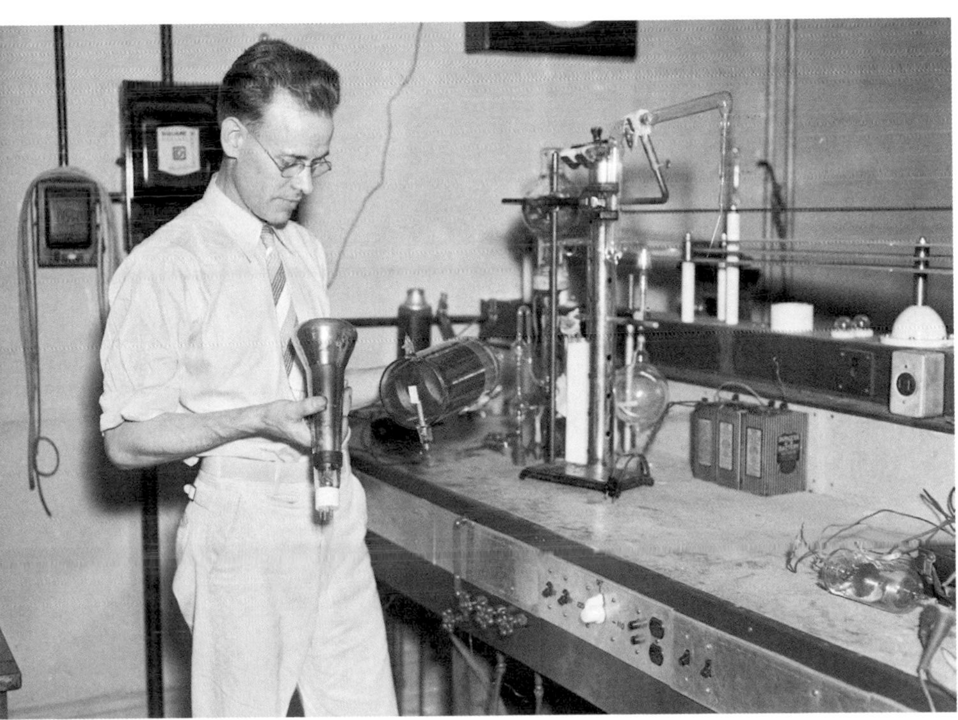

10. Often described in America as 'the farm boy who invented television' Philo Farnsworth is pictured here with his 'image dissector'. Though he had only minimal scientific training, Farnsworth had an ambition to be the first to invent 'all electronic' television. He had considerable success but failed in the end to beat the big corporations who he accused of stealing his technology.

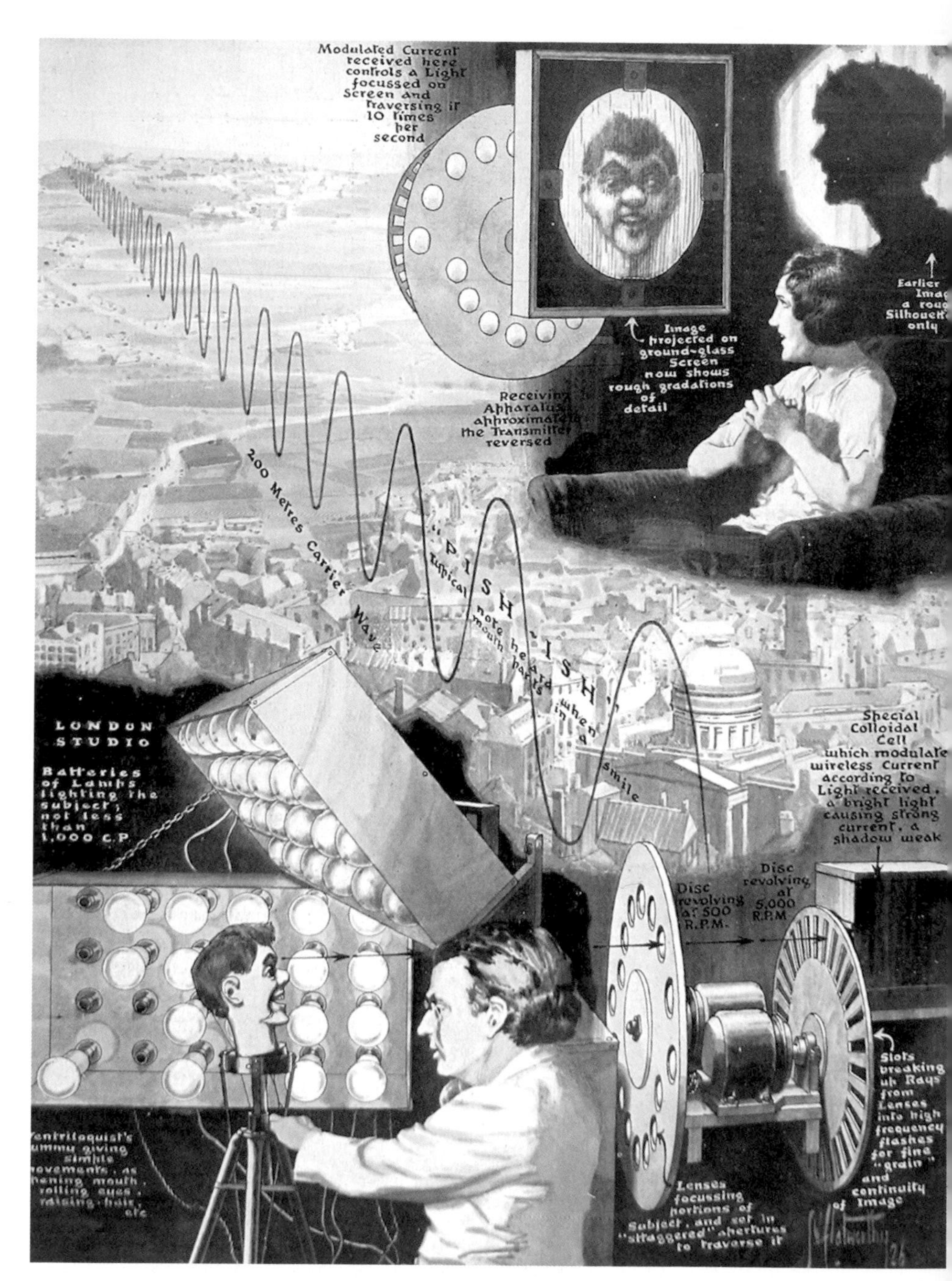

11. A graphic artist's attempt to illustrate the working of John Logie Baird's semi-mechanical television system. Published in *The Sphere* in 1926, a year after Baird first managed to transmit an image in his workshop in Soho, the newspaper covered the story of a broadcast from London to Harrow nine miles away. The image being sent is of 'Stooky Bill'.

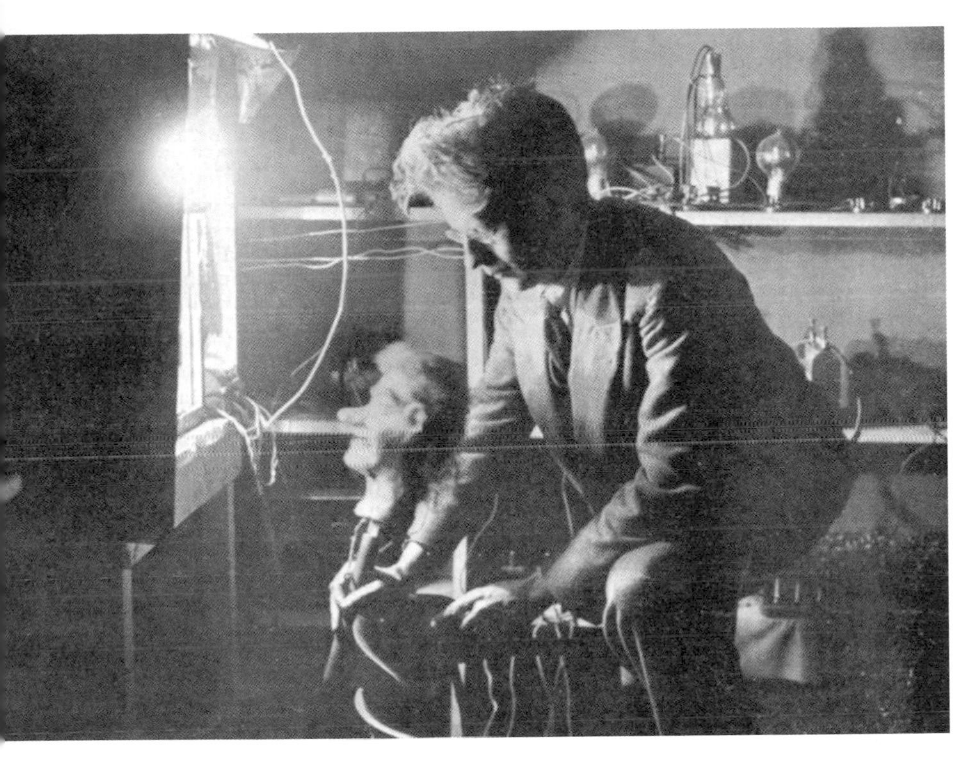

12. Baird's ramshackle television technology required a very bright, intense light to register an image and in the early days only a dummy's head could withstand the heat. Stooky Bill, as the dummy was christened, played his part. Baird realised he had made a breakthrough when the features of Stooky's face appeared on screen for the first time in October 1925. This was Baird's eureka moment.

13. Born in Russia in 1889, Vladimir Zworykin worked as a student with the scientist Boris Rosing on a television system using a cathode ray tube, which was patented in 1907. The outbreak of war in 1914 and the Russian Revolution put an end to what had been an idyllic life and Zworykin escaped to America in 1919. He took out a patent for a television system in 1923 but nobody was interested then and it was in the 1930s that he produced his iconoscope, an electronic scanner. Later he said American television was 'awful'.

14. Joe Woodland, the man who dreamed up the bar code while sitting on a Miami beach in the winter of 1948–49. The crude system he rigged up with a friend was patented, but the technology was not there to make it commercially viable. It was not until the laser and the microchip arrived in the 1970s that the supermarkets took it on and revolutionized the check-out counter.

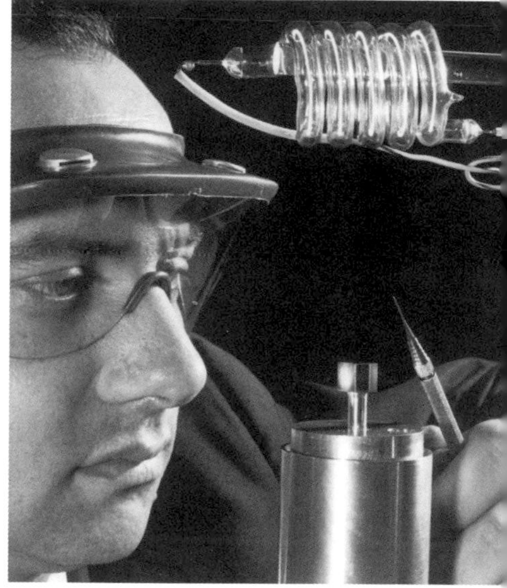

15. A theoretical paper by Albert Einstein in 1917, entitled *On the Quantum Theory of Radiation*, suggested that it would be possible to stimulate the emission of electromagnetic radiation. Scientists proved the theory to be correct by first producing the maser, which amplified microwaves, and there was a race to produce the laser, that is, 'light amplification by the stimulated emission of radiation.' In 1960 Theodore Maiman of the Hughes Aircraft Company, pictured, was the first to demonstrate the powerful light of a laser which many journalists dubbed the 'death ray'. Maiman did not anticipate the use of lasers for reading bar codes.

16. Before the bar code appeared in supermarkets a system called KarTrak had been used on the railroads in America to identify and locate freight cars, which had a habit of getting lost. Devised by David Collins and a team at the Sylvania electronics company, coloured bar codes on the side of the cars were read by a powerful light. Adopted in 1967 by the American Association of Railroads 1.5 million cars were labeled. Pictured here is a bar code and scanner at Midland, Michigan in 1968. KarTrak was abandoned later because of poor maintenance of the system.

NCR 255 scanning system for supermarkets extends computer's power to checkstand. First system installed in U.S. is in Marsh Super Market, Troy, Ohio. Checker passes purchased items over scanning window. Universal Product Code, which appears on package, is read by laser scanner linked to computer. The latter records items and flashes prices on display panel. In supermarket control room, NCR 726 minicomputer controls system and provides detailed operating information for store manager.

17. An historic check-out in the Marsh supermarket in Troy, Ohio where, on 26 June 1974, the first item marked with a Universal Product Code, agreed to by the US grocery trade, was scanned. Clyde Dawson, head of research and development at the store, chose as the first item a multi-pack of Wrigley's Juicy Fruit chewing gum as he was impressed that a bar code would fit onto such a small item. The occasion is still celebrated every few years in Troy.

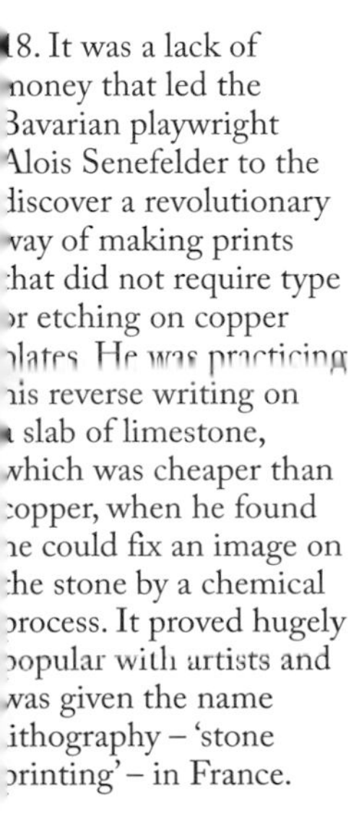

18. It was a lack of money that led the Bavarian playwright Alois Senefelder to the discover a revolutionary way of making prints that did not require type or etching on copper plates. He was practicing his reverse writing on a slab of limestone, which was cheaper than copper, when he found he could fix an image on the stone by a chemical process. It proved hugely popular with artists and was given the name lithography – 'stone printing' – in France.

19. A hand operated lithographic press invented by Alois Senefelder. It was lithography that inspired Niépce in France and Fox Talbot in England to experiment with the chemical fixing of images using light, which was at first called heliography or 'sun drawing'.

20. Nobody in London took any interest in the pioneering photographs taken by Niépce, but Louis Daguerre in Paris recognised their potential and went into partnership with him. When Niépce died, Daguerre, pictured here, stole his thunder and renamed the images after himself. The *Daguerreotype* became hugely popular. By the 1870s microphotography led to the miniaturization of images, which was ultimately used in the production of microchips.

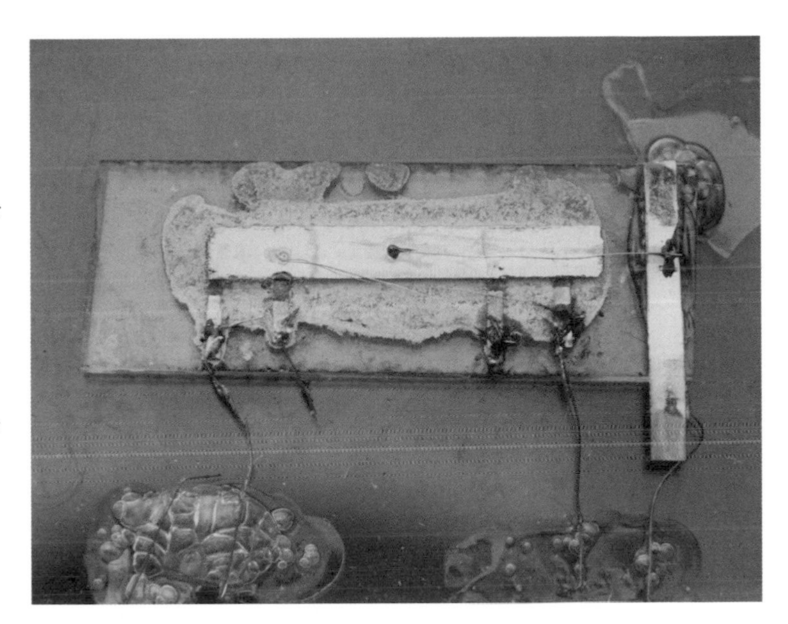

1. The original
ntegrated' circuit
reated by Jack
ilby in 1958
hen he was
orking for Texas
nstruments and for
hich he was later
warded the Nobel
rize in Physics.
Crude though it
as this was the
reakthrough in
he creation of
niniature computer
omponents.

22. One of the stars of Silicon Valley, Robert Noyce with the 'motherboard' of his version of the integrated circuit, which he developed independently of Jack Kilby. Noyce was a founder of Intel, the company that produced the chip, which inspired the creation of the first home computer kit.

3. It was a small firm in
Albuquerque, New Mexico
hat, in 1975, produced a cheap
ome kit for enthusiasts called
he Altair 8800 and began the
ersonal computer revolution.
Ed Roberts, pictured, had the
nspiration of using Intel's
nicroprocessor as the 'brain' of
low cost computer and made
imself a fortune. However, he
uit the computer business to
ulfill his boyhood ambition
o be a doctor. He ran a family
ractice until his death in 2010.

24. It was this issue of *Popular Electronics* in January 1975 that inspired Bill Gates and Paul Allen to write a programme for the Altair 8800, which Allen took to Ed Roberts in Albuquerque. It worked: the first calculation the Altair performed was 2+2=4. It was there that Microsoft began.

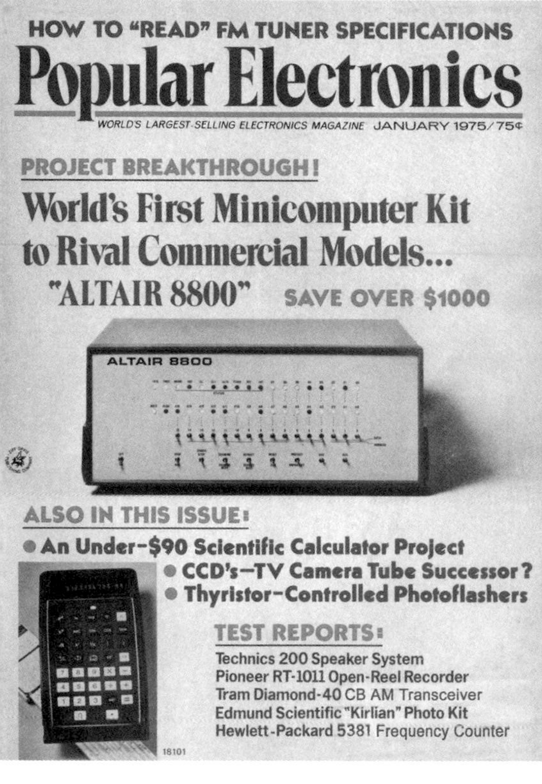

NEWSLETTER

Homebrew Computer Club

Robert Reiling, Editor – Post Office Box 636, Mountain View, CA 94042 – Joel Miller, Staff Writer
Typesetting, graphics and editorial services donated by LAUREL PUBLICATIONS, 17235 Laurel Rd., Los Gatos, CA 95030 (408) 353-3609

random data

by robert reiling

Personal and home computing continue to interest more and more people. The hobbyist clubs are expanding, more computer stores are opening, new products are being announced, magazines devoted to personal computers are increasing in number, and predictions are being made about where we are going. A few people are attempting to determine what has happened in the past two years and how many computers and computer people are out there.

This issue of the NEWSLETTER has a survey that presents a list of the Homebrew Computer Club systems. I even projected that data. Maybe you would like some facts about the Homebrew Computer Club, not projections just facts. OK, from the history file:

Homebrew Computer Club first meeting–March 5, 1975
First meeting attendance–32 people
First newsletter published–March 15, 1975
Homebrew Computer Club meeting–October 15, 1975
Attendance–80 people
Computers up and running–38

Homebrew Computer Club Newsletter distribution March 1976–600 copies

Homebrew Computer Club meeting–June 9, 1976
Attendance–250 people
Computers up and running–101

Homebrew Computer Club meeting–January 19, 1977
Attendance–240 people
Computers up and running–182

Homebrew Computer Club Newsletter distribution January 1977–1500 copies

According to Gordon French, the Homebrew Computer Club is the oldest major hobbyist club in the world. Gordon's garage was the location of the first meeting.

The *Wall Street Journal* published a front page article Friday, February 4, 1977 titled: Home Input, TheComputer Moves From the Corporation To Your Living Room. The article by David Gumpert, staff reporter of *The Wall Street Journal*, makes this observation: "The home-computer industry is so new and so fragmented that it hasn't got around to computing its own progress, so nobody knows how many individuals have bought computers. But estimates range from 20,000 to 100,000." There is much more in this article including some quotations from Jim Warren, a member of the Homebrew Computer Club and editor of Dr. Dobbs Journal. Look this article up if you missed it.

I note that De Anza College is offering microcomputer programming courses. Contact DE ANZA COLLEGE – SHORT COURSES, 21250 Stevens Creek Boulevard, Cupertino, CA 95014.

David Bunnell, publisher of Personal Computing Magazine, has announced the acquisition of Microtrek Magazine which will become a special section in Personal Computing. Microtrek, the computer magazine for the hobbyist, was first published in August of 1976. Its second edition was recently published in December of 1976. Subscribers to Microtrek Magazine will begin receiving Personal Computing Magazine with the upcoming March-April edition.

Apple Computer, Inc., 770 Welch Road, Palo Alto, CA 94304, telephone (415) 326-4248, has advance order information for the Apple-II. The Apple-II consists of a 6502 microprocessor, video display electronics including color graphics, RAM, ROM, ASCII keyboard port etc. all on a single PC board. If you order now, delivery is expected to be no later than April 30, 1977.

Don't forget articles are needed for the NEWSLETTER. Also, don't forget your donations are needed to pay for postage, printing, etc. □

HCC Newsletter/Vol. 2, Issue 14/February 16, 1977 *one*

25. The cover story of *Popular Electronics* advertising the Altair 8800 was greeted with great excitement by the computer geeks in California. The Homebrew Computer Club was formed in March 1975 to explore ways in which the basic kit could be improved upon and held its first meeting in a garage. One of those who went along was Steve Wozniak who knew nothing about the Altair 8800 but realised its potential.

6. Founders of Apple, Steve Wozniak, on the left, and his friend Steve Jobs with some of their arly computer equipment. Wozniak was the electronics wizard who realised that with the echnology used to build the Altair 8800 he could fashion something much more sophisticated. obs was the businessman who suggested they form a company reasoning that if it failed they could t least say they did once own a business.

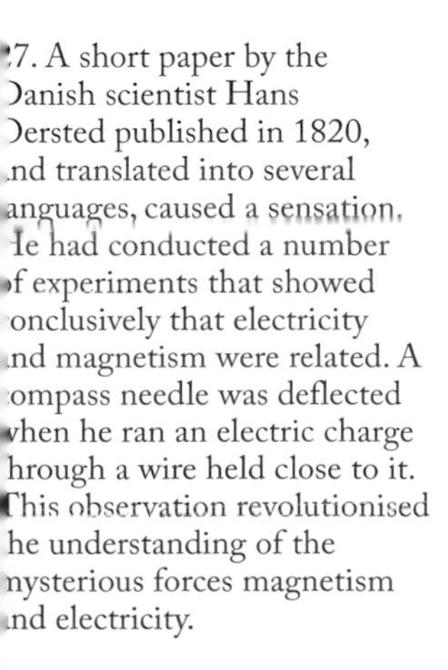

7. A short paper by the Danish scientist Hans Dersted published in 1820, nd translated into several anguages, caused a sensation. He had conducted a number of experiments that showed onclusively that electricity nd magnetism were related. A ompass needle was deflected vhen he ran an electric charge hrough a wire held close to it. This observation revolutionised he understanding of the nysterious forces magnetism nd electricity.

28. Michael Faraday, the man who discovered how to generate electricity, wit his wife Sarah. Both were members of the Protestant sect called Sandemanians and Faraday remained true to his faith while establishing himself as one of the most revered scientists of his day. He had begun work as an apprentice bookbinder, work h hated, before working as an assistant to Si Humphry Davy at the Royal Institution.

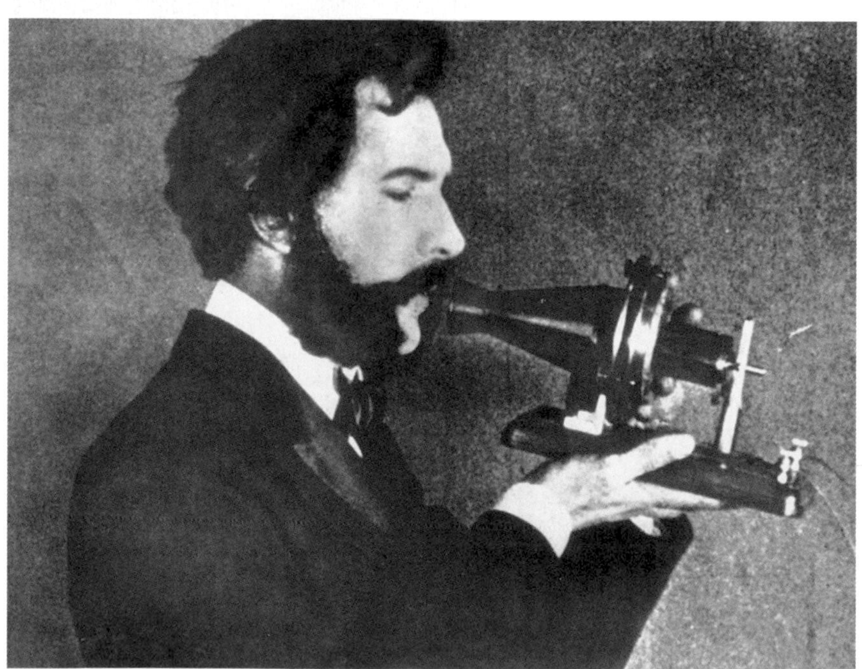

29. Alexander Graham Bell with his newly invented telephone. Born in Edinburgh in 1843 Bell came from a family of speech therapists with a special interest in helping deaf children to speak. In fact Bell always described himself as a teacher of the deaf rather than a scientist and he had little electrical knowledge. He patented his invention in 1876, seven years after he had emigrated with his parents to Canada and then North America. A year later he married Mabel Hubbard who had been deaf from the age of five. Shares in Bell's company made them instantly wealthy.

0. It was Guglielmo
/Iarconi, the dapper young
talian who spoke perfect
English, who invented the
irst commercial wireless
elegraph in the 1890s in
England with financial
acking. He is seen here
vith his loyal assistant
George Kemp who he
ad met demonstrating
is system to the General
Post Office. Only the dots
nd dashes of Morse Code
ould be transmitted with
is pioneer wireless but
/Iarconi became famous
vhen his telegraphers
layed a part in dramatic
escues at sea. The
urvivors of the Titanic
lisaster in 1912 owed
heir lives to messages
apped out by the Marconi
perators on board.

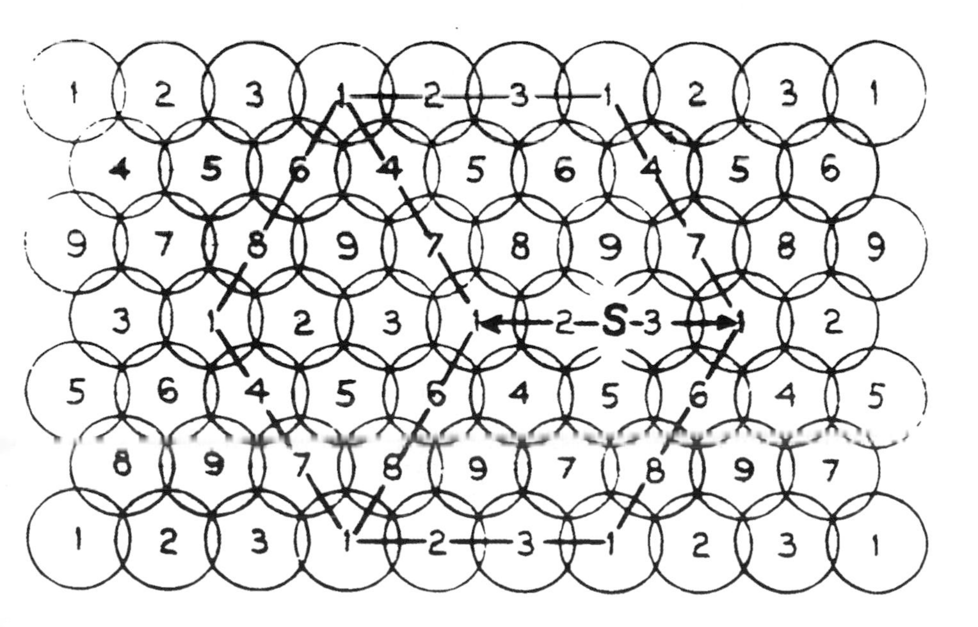

31. The basic concept of a 'cellphone' system sketched by Doug Ring of Bell Telephone Laboratories n 1947. To avoid interference between the mobile phone users signals are switched from one cell o another with a technique known as 'handover'. Ring's concept remained dormant until the rrival of the microcomputer in the 1970s.

32. Mobile phones in the 1950s were car phones, like this one made by General Electric. The car boot was full of wireless equipment and callers were connected by operators. Only a handful of subscribers could dial in at the same time. With its honeycomb of cells working on different wavelengths and connecting to base station the Bell system would eventually overcome this problen

33. In 1973, Motorola, a rival of Bell Telephones, had a great publicity coup with the launch of the first genuinely hand-held mobile phone, which it calle the DynaTAC. Vice President Martin Cooper had the job of commissioning the hand-set, soon nicknamed 'the brick', and is pictured here making one of the first phone calls. A brilliant piece of engineering, the DynaTAC was ahead of its time as, to operate as a genuine mobile, it had to await the establishment of a cellphone system.

The system spread to the US Postal Service and the sea container business. However, the pioneer scheme on the railroads fell into disuse after a few years because companies failed to keep the bar codes clean enough to read. Eventually KarTrak was abandoned and new identification systems were agreed upon. Meanwhile, David Collins had moved on.

Sylvania engineers had experimented with laser scanners and Collins and others were keen to develop a black and white bar code technology. But the company was happy to stay with the KarTrak system while it was still making money. In 1968 Collins left Sylvania and founded his own company, Computer Identics Corp., which was soon developing a bar code system for Buick Motors, reputedly the first to use the laser scanner to read a bar code. A variety of components engineered for Buick cars were barcoded so that as they left the factory on a conveyor belt the scanner logged them providing hourly reports of what had been despatched.

Bar codes were making inroads. But they had yet to make much of an impression on the grocery industry which had first inspired Joe Woodland to draw his fingers through the sand on Miami Beach. There were three good reasons for this. First, the laser and computer technology had to become miniaturised and relatively inexpensive, as installing bar code systems would be a huge expense. Secondly, the problem of labelling hundreds of thousands of products presented a challenge for the printing of labels and inventing of a code that would be accepted by manufacturers and retailers alike. Thirdly, and most critically, the vast grocery industry would have to find a way of agreeing on a universally acceptable code and technology. The way in which this last problem was solved would make a fine Hollywood movie, as executives held brainstorming sessions across the continent while the engineers and scientists did their mathematical calculations and experimented with their 'symbologies'. The fulfilment of this epic enterprise was the creation of the UPC – the Universal Product Code.

* * *

A booklet produced in 1966 by the Kroger Company, which ran one of the largest supermarket chains in North America, signed off with a despairing wish for a better future: 'Just dreaming a little ... could an optical scanner read the price and total the sale. ... Faster service, more productive service is needed desperately. We solicit your help.' Kroger's business was groceries, not electronics, so the company went looking for a partner with the necessary expertise. For a while they had discussions with Sylvania Electronics to see if they could produce an in-store version of the KarTrak system. By that time Sylvania had lasers and could almost certainly have found a way to make such a system work, but they chose not to. Like many other electronics firms they did not regard the supermarket business as potentially lucrative. However, Sylvania by chance led Kroger to the huge and powerful Radio Corporation of America (RCA) which was diversifying and looking for new lines of business.

'It was just one of those days in the mid–60s at the RCA Missile and Surface Radar division when everyone seemed to be waiting and preparing for the award of the next big defense contract,' Francis Beck, Technical Director of the RCA Princeton Laboratory, recalled in an interview for the online ID History Museum. While rumours of the big contracts were being exchanged, Beck learned of a small group being put together to explore new ways in which RCA could exploit its expertise. 'Was it possible that I could pursue my entrepreneurial fantasies under an umbrella of security offered by a large and successful corporation? I couldn't resist the temptation.'

Beck gave up radar design to turn inventor. The small research team looked at a few projects, including the possibility of an automatic bank cash machine, which they decided would not go because 'the customer would not buy the concept'. Finally they lighted on the bar code. A search of the history turned up some apparently hare-brained schemes: in one, customers picked out punch cards which identified what they wanted to buy and presented them to a cashier, who retrieved the goods from a store. This did not survive long in the grocery business. Then there was the patent for a system in which the supermarket shopper threw everything into a basket, which was pushed under a scanner that identified each item and printed out a bill. This remained on the drawing board.

They soon found the Woodland and Silver patent which the owners, Philco, had attempted to put into practice. It failed because they could not find a way of printing the bullseye bar code satisfactorily. This was not the rectangular bar code that Woodland had first envisaged on Miami Beach but the 'bullseye' of concentric circles he thought would be a better design. As he recalled, having elongated the dots and dashes of Morse Code he swept his fingers around in the sand. When he and Bob Silver worked on it they decided the bullseye was the better symbol because it could be read accurately from any angle.

The Woodland–Silver bullseye patent ended up with RCA while Joe Woodland was working at IBM. Working closely with the Kroger Company, Francis Beck and the bar code technicians began to fashion a working system. The dream of the supermarket boss began to turn into reality. With the laser, and new techniques for printing, the technology was almost keeping pace with the imagined automatic checkout system.

Computers were a real problem at the outset. At least ten stores would have to share one central computer. The costs of implementing any scheme looked daunting. Even in 1968 the trade journal *Supermarket News* had a headline saying that the heavy investment costs led many to believe 'front end automation' would not be feasible. There was also a worry about lasers. Would the 'death ray' tag frighten customers away?

The descriptions given by engineers such as Francis Beck of their first demonstrations of a prototype bar code checkout system, though deadly serious for them, are the stuff of light comedy for the rest of us. Charlie Chaplin, who made such fun of new technologies, would have captured the essential absurdity. A group of high-powered executives from the supermarket chain gather around a mock-up of a store checkout. Engineer Beck is detailed to demonstrate the system. Robert Sarnoff, chairman of the RCA Board, plays the part of the customer. He hands Beck a 12-ounce jar of Acme Ideal brand smooth creamy peanut butter, identified by a bullseye symbol printed on the bottom. The symbol has been deliberately damaged to add a touch of realism to the otherwise fantastical scene. Beck, playing the checker, swipes the peanut butter jar across the scanner and the price is displayed. According to Beck, Sarnoff appeared to be 'fascinated' by this automated checkout, if not entirely convinced that this was the future of supermarket shopping.

Printing the bullseye bar code proved to be one of the greatest difficulties, because any imperfections would make the whole system unworkable. A rotating turret of ballpoint pens, and a pen designed for astronauts which could write upside down, solved some of the problems. All this technical development, involving several companies commissioned by RCA, was to lead up to the first real life test at the Kroger Kenwood Plaza store in Cincinnati. On 3 July 1972 the first automated checkstands were installed. Shoppers were offered compensatory stamps if the price displayed was not the same as that on the product. Gradually, price labels were removed. More checkstands were installed and a comparison with other Kroger stores told an undeniable and very promising story: the bullseye bar code hit the target, with superior sales figures. But this was just one store in a nationwide grocery and supermarket business worth billions. If the laser and bar code were to revolutionise the checkout counter they would have to be near universal. It was the dilemma that the railroad companies had faced with KarTrak: it was all or nothing. As it turned out, RCA's pioneer checkstands were already history: one of them is preserved in the Smithsonian. But they played a vital part in the development of the bar code on a national and international scale.

* * *

Francis Beck had been anxiously watching customer reactions to the novel checkout system at the Kenwood stores for just over a month when a high-powered delegation representing major supermarket chains and grocery manufacturers paid a ceremonial visit. This was the wonderfully titled Symbol Selection Committee chaired by Alan Haberman, executive vice chairman of First National Stores, and with expertise from Heinz, Del Monte, Procteor & Gamble and General Foods. With them was the lawyer Stephen A. Brown who would author the book *Revolution at the Checkout Counter: The Explosion of the Bar Code.* He wrote: 'On August 10 1972, the committee saw the future when it visited the Kenwood store.' It was the first time most of them had seen in action something which had been a figment of their imagination, a dream.

The committee was a technical offshoot of something called the Ad Hoc Committee of the Universal Product Identification Code. Nobody who observed the first meeting in 1970 of suited representatives of the grocery trade assembling in the grimly featureless O'Hare Inn in Chicago, with its view of the end of an airport runway, could have guessed that they were about to revolutionise not just the supermarket checkout all over the world but just about every other aspect of modern life which benefits from electronic tagging and identification. Members of the Ad Hoc Committee themselves were not confident they could pull it off, but they felt compelled to give it a try because the supermarket business was in trouble, with profit margins below 1 per cent. The technology had not kept pace with the expansion of the trade. Haberman told the *Smithsonian Magazine* years later:

> While the rest of the world was sending people to the moon and playing around with computers, nobody in the technological world wanted to work in the supermarket business, because it was such a low profit industry. The business was literally in the backwash of technology . . . the checkout experience was the least pleasant experience in the store. It was the thing people hated! They hated having to wait in line! They hated having to watch the checker's action. It was an unhappy place.

The goal of the Ad Hoc Committee could be stated very simply. It was charged with finding a way to introduce a Universal Product Code, a bar code of some description – what Haberman liked to call a 'footprint' – which would be common to all goods sold in supermarkets and imprinted by the manufacturers and retailers. The code would carry information about the nature of the product, the company that made it, and so on. In-store computers would 'read' this information with scanners and introduce their own variations, which might involve special offers and reductions. The vision was there but the difficulties in the way of its realisation were daunting. One of the first steps the committee took was to appoint McKinsey & Co. as consultant to manage the project.

Manufacturers were often resistant to the idea of a universal code. They had existing methods of identification of products, which would have to be discarded or adapted. Cardboard manufacturers worried that a printed code might spoil their product. Canners did not want to be obliged to put bar codes on the base of cans. As the quest of the committee became known the electronics industry began to take notice. It was the task of the Symbol Selection Committee to review rival technologies but they were hampered by the natural reluctance of competing companies to reveal their secrets.

It took four years to arrive at a workable proposition to put to the whole industry. When there was broad political agreement on the desirability of a Universal Bar Code and the feasibility of printing it on a huge variety of products at the point of production as well as in stores, the work of the Symbol Committee focused on the bar code itself and how it could best be read and linked with in-store computers. Their research was not confined to the United States; several committee members went on a European tour. In the end seven companies, all of them based in the United States, submitted systems to the committee. One of these, Charecogn Systems, founded in 1960 by A. John Esserian, had demonstrated to newspapers what would have been a fully functioning checkout system as early as August 1970. Esserian liked to call this the 'Kitty Hawk of Supermarket Bar Code Scanning' but it was developed for the United States Department of Agriculture rather than the grocery trade.

RCA, having demonstrated to the committee its system in Cincinnati, took the view, not unreasonably, that it was the only real contender.

National Cash Registers (NCR) also had its own bar code system in operation in a small way. However, at the last minute, a surprise bid was made by International Business Machines (IBM). It had no technology at all to demonstrate to the committee and the decision to enter the competition appears to have been an afterthought, despite the fact that it had in its employ none other than Joe Woodland. As it turned out, although he was involved in IBM's submission, he was not the creator of its version of the Universal Bar Code. That fell to George Laurer who, in his own view, had an advantage over his rivals because neither he nor IBM had given supermarket checkout systems or bar codes much thought and his company had no ready-made technology. Starting from scratch, Laurer had no prejudices about the appearance of the bar code, though his bosses had assumed it would be some version of the circular bullseye in Woodland's patent and RCA's pioneer system in Cincinnati.

In his book *Engineering Was Fun* Laurer tells the story:

> In the fall of 1971, Paul McEnroe (my manager at the time) gave me a letter and an assignment. The letter was from our division headquarters directing him to prepare a presentation for the division dignitaries and the head of our laboratory, supporting RCA's bull's-eye code19 and symbol. He handed it to me and said do it because he was leaving on vacation and would not be back until the day before the scheduled presentation. – GOODBYE!
>
> I struggled a day or two but my nature and training would not allow me to support something I did not believe in. It was obvious to me that that approach [the bullseye symbol] would never satisfy all the requirements over the long run, even though RCA was demonstrating their system at the time. I simply went against my manager's instruction and set out to design a better system.

Laurer was handed the specifications for a bar code that had been determined by the Symbol Selection Committee: it had to be small and neat, maximum 1.5 square inches; to save money it had to be printable with existing technology used for standard labels; it had been calculated that only ten digits were needed; the bar code had to be readable from any direction and at speed; there must be fewer than one in 20,000 undetected errors. Laurer recalled:

> The first break came with the realization that if one simply deflected a single scanning line with a 'corner' mirror – two mirrors at right angles – an 'X' would be formed on the scanning window. With a simple 'X' linear bars could be read no matter how they were oriented in the scan window. I devised a six module code and incorporated a rudimentary parity check in the start and stop bars of the symbol.
>
> I thought it would work, it was not great – but it certainly had a better chance of success than the bull's-eye. I prepared a presentation, which explained why the bull's-eye was not the correct choice and offered my solution instead. I finished the presentation and charts, with the help of my 15 year old son Craig on a Sunday, the day before the scheduled presentation. My manager lived across the street from me so when I saw him arrive home from his vacation, I went over to his place with the charts and explained what I had done. He had no choice but to agree but he also made it clear that if I was wrong or if I could not sell the idea to the brass it would end my career, not his. I was truly playing 'bet your job' by designing a new code and symbol rather than supporting what the brass wanted.

Although there was scepticism in IBM, Laurer was convincing enough to be given the go-head with his rectangular bar code. There were many adjustments to be made and the Symbol Selection Committee changed the brief on occasion. The time came when he would have to convince his boss Mr B.O. Evans that he had a winning formula. Everything up to that point had been theoretical. A division of IBM built a prototype scanner and Laurer's Universal Product Code was tested. 'There were many sceptics in IBM,' Laurer recalled, 'not the least of whom was B.O. Evans himself. However at the end of a flawless demonstration for Mr Evans, we had our ace softball pitcher pitch bean bag ash trays, with symbols on the bottom, as fast as he could over the scanner. When each one read correctly, Mr Evans was convinced.'

It was another matter to convince the Symbol Selection Committee, which was under huge pressure to accept RCA's already functioning bullseye symbol and technology that had done much to inspire confidence that a universal product code could work. In fact RCA gave the committee an ultimatum: either its symbol was selected or it would pull out of the market. Stephen Brown put it starkly: 'They essentially were presenting us with "You play with our marbles or we're going home."' However, IBM put in a strong challenge. Brown recalled: 'As IBM's presentation unfolded, Eric Waldbaum [representing Greenbelt Consumer Services] turned to me and whispered 'My God, they have an entire system developed, and we didn't know anything about it.' At one point, the IBM spokesman reached into his back pocket and pulled out a silicon wafer about the size of a silver dollar. This wafer, he said, contains all the computing power necessary to run a checkout system. The IBM explication was a tour de force...'

By March 1973, the members of the Symbol Selection Committee agreed that it was time they chose the form of bar code that would be the blueprint for a universal product code for the industry. It was a huge responsibility as it would involve both manufacturers and retailers in enormous expense. The technology of the bar code and its relationship to a computer program was still quite new. George Laurer at IBM had argued forcefully that the bullseye bar code would fail because of the problems of printing it. At one of the committee's meetings in San Francisco the chairman, Alan Haberman, sought to reduce tension by taking them first to a popular restaurant to wind down, and then to a showing of the pornographic film *Deep Throat* starring Linda Lovelace. There is no discernible influence of the film on the shape of the printed code.

After asking for an appraisal of the rival symbologies from scientists at the Massachusetts Institute of Technology on 30 March 1973, in a New York hotel close to Grand Central Station, the committee met to make its final and fateful decision. Haberman asked them first to declare how sure they were that the symbol they had chosen was the correct one. There was a very high level of confidence – about 90 per cent all round – and the winner was Laurer's rectangular code, which had undergone considerable modification since he first proposed it. For Woodland, who died in 2012 at the age of ninety-one, it must have been a strange experience to witness the reincarnation in sophisticated form of the elongated lines of Morse Code he had drawn in the sand of Miami Beach in 1949. There was now a modestly priced laser scanner to register with a concentrated beam of light the coded vertical lines of alternating black and blank and a microcomputer to decipher the information.

* * *

Every few years, the small town of Troy in Miami County, Ohio celebrates an historic occasion that for a few giddy weeks puts it on the world map of the grocery trade. At the time National Cash Register, which provided the checkout equipment, was based in Ohio and Troy was also the headquarters of the Hobart Corporation, which developed the weighing and pricing machines for loose items such as meat. It was here, at just after 8 a.m. on 26 June 1974, that the first item marked with the Universal Product Code (UPC) was scanned at the checkout of Troy's Marsh Supermarket. It was treated ceremonial occasion and involved a little bit of ritual. The night before, a team of Marsh staff had moved in to put bar codes on hundreds of items in the store while National Cash Register installed their scanners and computers. The first 'shopper' was Clyde Dawson, who was head of research and development for Marsh Supermarket; the pioneer cashier who 'served' him Sharon Buchanan. Legend has it that Dawson dipped into his shopping basket and pulled out a multi-pack of Wrigley's Juicy Fruit chewing gum. Dawson explained later that this was not a lucky dip: he chose it because nobody had been sure that a bar code could be printed on something as small as a pack of chewing gum and Wrigley had found a solution to the problem. Their ample reward was a place in American history.

According to those who recall the historic occasion it was possibly that very pack, and not some equivalent, that was collected by the

Museum of American History at the Smithsonian, to be put on display there at a special exhibition. If that is the case, then a second, identical, pack was preserved by Marsh Supermarkets for their ritual celebration of the UPC and it has been brought out of the cupboard a number of times over the years. The fortieth anniversary was celebrated in July 2014.

The Ad Hoc Committee's view that shoppers would not have any concerns about the new checkout system proved to be optimistic. Labels were taken off individual items so that when the customers got to the till they did not know if the price rung up with the bar code was the same as they had seen on the shelf. To assuage their fears, in the first weeks they were given crayons so they could note the shelf price on each item. This was abandoned, and before long prices were marked alongside the bar code.

Like so many inventions, the UPC was not an immediate success. In fact when in 1976 only fifty supermarkets in the United States had installed bar code checkouts, compared with a projected one thousand when the UPC was launched, *Business Week* ran a piece headlined 'The Supermarket Scanner that Failed'. Nobody doubted that the bar code would make running a store cheaper and more efficient, but the initial investment was daunting. It was when the mass merchandisers adopted the UPC that it took off, Kmart being the first. In fact bar code technology was almost made for companies like Walmart, which deal in thousands of goods that need to be catalogued and tracked. The bar code took off in the grocery and retail business in the 1980s, and at the same time began to transform manufacturing and to appear like a rash on anything that benefited from instant identification. In 2004, *Fortune* magazine estimated that the bar code was used by 80–90 per cent of the top 500 companies in the United States.

Though the inspiration for the bar code was the plea by supermarkets for technology that would speed up the checkout, its greatest value to business and industry is that it has provided hard, statistical evidence for what sells and what does not. It has transformed market research, providing a rich picture of people's tastes, and it has made production lines more efficient. The once-dreaded 'death ray' laser beam now comes in handy gun-sized scanners which instantly read and log anything from hospital drugs to newborn babies. LED (light-emitting diode) lights are also used to provide the beam for scanners. The UPC spawned an entire bar code industry which now has well over 200 distinct symbologies tailor made for different industries and specialisms.

After many years of anonymity, the man whose knowledge of Morse Code inspired the familiar black and white stripes finally got some recognition. The circumstances were, as with so much in the bar code's history, a little comical. In February 1992, President George Bush Snr was photographed at a national grocery convention looking intently at a supermarket scanner and having a go at swiping a can with a bar code over it. The *New York Times* correspondent wrote this up as evidence that it was the first time Bush had seen a supermarket checkout: in other words, he was out of touch with everyday American life. His aides insisted that he was not struck by the novelty of the technology but by the fact that it could read a damaged bar code. Apocryphal or not, the story stuck and was regarded as damaging to Bush. However, as Woodland's local newspaper put it: 'George Bush isn't one to hold a grudge. No Sir.' A few months after the checkout incident Bush presented Woodland with a National Medal of Technology. Finally Joe had got recognition, forty-four years after he had first dreamed up the bar code on Miami Beach.

It was not only laser technology that made the bar code a practical possibility. Just as important was the invention of the microprocessor, which revolutionised so much of modern life in industry and in the home. It made possible the personal computer which, even in the early 1970s, had been the dream of only a few hackers and geeks. As so often in the history of innovation, it was these enthusiastic amateurs who first learned how to turn the astonishing new technology into familiar domestic appliances. 'Triumph of the Nerds' was the title of a British–American documentary which told the story in 1996.

CHAPTER 4

HOMEBREWED

'The Home Computer is Here,' the American magazine *Popular Electronics* announced to its readers in the January issue of 1975. 'For many years we've been reading and hearing about how computers will one day be a household item. Therefore, we're especially proud to present in this issue the first *commercial type* of minicomputer project ever published that's priced within reach of many households – the *Altair 8800*, with an under $400 complete kit cost, including cabinet.' The Altair 8800 that was proudly displayed on the cover of *Popular Electronics* was, in fact, a mock-up, an empty shell, as the genuine prototype had been lost in transit between Albuquerque, in New Mexico, where it had been manufactured, and the magazine's offices in New York. Promoting it was an act of faith and a desperate attempt by the publishers to outdo their great rival, *Radio Electronics*. And it worked in a spectacular fashion, beyond their wildest dreams.

The tiny company in Albuquerque which made the Altair 8800 was very close to bankruptcy. But this minicomputer revealed a massive pent-up demand among electronics hobbyists in America. Such was the clamour to get hold of the Altair 8800 kit that some drove to Albuquerque and parked outside the makeshift workshop in a shopping centre and waited for their order to be delivered. This desire to own a 'home computer' was remarkable for a number of reasons, not least of which was the fact that the kit, as delivered for $397, was very nearly useless. Rows of flashing LED lights on its front panel indicated when it was working. However, as Steven Levy put it in his book *Hackers: Heroes of the Computer Revolution*: 'For all practical purposes, it was deaf, dumb and blind. But, like a totally paralysed person whose brain was alive, its non-communicative shell obscured the fact that a computer brain was alive and ticking inside. It was a computer and what hackers could do with it would be limited only by their own imaginations.'

Computer geeks, frustrated by years of queuing up for computer time and sneaking into offices at night to tap the mainframes guarded by what they called 'the priesthood', did not care that the Altair had no keyboard, no screen, no mouse and no program: just a panel of winking lights. However, the man who had devised it, Ed Roberts, had made sure that with some add-ons the Altair would operate like a real computer. What he was selling was the brain. And it was not long before help arrived to get the nascent home computer thinking and talking.

The *Popular Electronics* feature caught the imagination of two young men who, since their schooldays together in Seattle, had been computer geeks. Bill Gates and Paul Allen had learned programming when they volunteered with other boys to debug the computer of a local firm. They did it for nothing at first, and then for pay. At the time they saw the Altair feature in *Popular Electronics* Bill was a freshman at Harvard and Paul was working for Honeywell in Boston. They knew instantly that this minicomputer needed a language that could be handled by the microprocessor which was at its core. And they set out to rework the standard language known as BASIC: Beginner's All-Purpose Symbolic Instruction Code.

It did not matter that they did not have an Altair: they got hold of the manuals and were able to get the specifications for the microprocessor made by a company called Intel. And they had access to a computer at Harvard that would allow them to simulate the workings of the Altair. They phoned Ed Roberts and offered to deliver a version of BASIC which could run on the Altair and which could be licensed to sell with it. When they had it ready to go Paul Allen flew out to Albuquerque with the program on ticker tape. He finished some of the work on the plane. 'I came out of the terminal sweating and dressed in my professional best, a tan Ultrasuede jacket and tie,' Allen wrote in his autobiography, *Idea Man*. 'Ed Roberts was supposed to pick me up, so I stood for about ten minutes looking for someone in a business suit. Not far down the entryway to the airport, a pickup truck pulled up and a big, burly, jowly guy – six foot four, maybe 280 pounds – climbed out. He had on jeans and a short sleeve shirt with a string tie, the first one I'd seen outside of a Western. . . . As we bounced over the city's sun-baked streets, I wondered how all this was going to turn out.'

Paul Allen spent an anxious night in the Sheraton Hotel: Ed had to foot the bill. The next morning would be the moment of truth when he fed the BASIC into the Altair at the workshop of MITS, Ed's company. It was tense for all of them, not just for Allen but for Ed and his engineer Bill Yates too. When he had got the Altair primed and the BASIC fed in, Allen hit the 'run' button. He recalled: 'The Teletype's printer clattered to life. I gawked at the uppercase characters: I couldn't believe it. But there it was: MEMORY SIZE?'

This was certainly a 'eureka moment' for all of them. Allen recalled:

> 'Hey,' said Bill Yates. 'It printed something!' It was the first time he or Ed had seen the Altair do anything beyond a small memory test. They were flabbergasted. I was dumbfounded. We all gaped at the machine for a few seconds, and then I typed in the total number of bytes in the seven memory cards: 7168.

The Altair responded with 'OK'. But there were still tests to do before Allen finally asked the computer to perform a simple addition: 2 + 2. The answer 4 was instant.

'That was a magical moment,' Allen wrote. 'Ed exclaimed 'Oh my God, it printed 4!'. . . . But Ed wasn't as surprised as I was that our 8080 BASIC had run perfectly the first time out of the chute. The Altair's one digit response, the classic kindergarten computation, proved that my simulator was on target. I was quietly ecstatic and deeply, deeply relieved.' Roberts straight away asked Paul Allen to work with them at MITS and Bill Gates later abandoned his studies at Harvard and joined his old school friend in Albuquerque. It was here, refining their BASIC for the Altair, that they created their company, Microsoft, which was to make them billionaires.

Ed Roberts was born on 13 September 1941 in Miami, Florida, where his father had an appliance repair business. He was interested in electronics as a schoolboy but decided he would like to become a doctor and enrolled as a medical student at the University of Miami. A neurosurgeon there suggested that, given the way medicine was going, Roberts ought to learn about electrical engineering. On this advice he switched courses. However, he left university abruptly when his wife became pregnant and he had to go to work to support his family. In May 1962 he enlisted in the US Air Force to take advantage of a scheme in which he would be paid to attend college. This took him in time to Lackland Air Force Base in San Antonio, Texas where he studied and then taught in the Cryptographic Equipment Maintenance School. To make some extra money Roberts, always a keen tinkerer and hobbyist, set up a company he called Reliance Engineering, which would take on any work going. This included the electronic animation of the Christmas characters in the window display of a famous San Antonio department store.

Roberts went back to college as a commissioned officer and graduated in electrical engineering from Oklahoma State University in 1968. From there he was assigned to the Laser Division of the Weapons Laboratory at Kirtland Air Force Base in Albuquerque, New Mexico, which is how the breakthrough personal computer came to be made there. At the Weapons Lab Roberts became friends with Forrest Mims who had returned from duty in Vietnam, and they found they had a common interest in model rocketry. With others they created a company they called Micro Instrumentation and Telemetry Systems, the acronym intentionally mimicking MIT, the Massachusetts Institute of Technology. MITS was established in Roberts's garage and began to market electronic kits for rocket enthusiasts. This was in the late 1960s when the Space Race was news, Albuquerque had a model rocket club and there was a publication called *Model Rocketry*, which thrived between 1968 and 1972.

Mims had ambitions to become a writer for science and electronics magazines and had his first piece published in *Model Rocketry*. Although he did not stay with Roberts in MITS it was Mims who introduced the company to Les Solomon, his technical editor at *Popular Electronics*, a flamboyant and influential figure in the world of technical publishing. Solomon, known as 'Uncle Sol' to the hobbyists, was always on the lookout for eye-catching and novel electronic kits which would attract the hobbyist readers. He visited Mims and Roberts in Albuquerque and they got on well. One of their successful collaborations was the publication in *Popular Electronics* of an inexpensive calculator kit, which Roberts developed. For a time it made him a good profit. But the rug was pulled from under him by the big electronics companies, which undercut him, selling calculators ready made for less than the MITS kits.

It was Roberts who had the inspiration that the microprocessors being produced by a young dynamic company called Intel were not just handy for calculators; they were sophisticated enough to power a small computer. It was his involvement with the hobbyists and *Popular Electronics* magazine which gave Roberts the idea that a computer kit might sell. There is no doubt that with the Altair 8800 he founded what was to become an entirely new and fabulously lucrative industry. It made Roberts a millionaire when he sold up and decided to return to his first love, medicine. He went back to college to qualify as a doctor and had a country practice for the rest of his working life.

It is impossible to exaggerate the influence of the Altair on the creation of the personal computer, but, like the first stage of a rocket, it fell out of sight after its spectacular launch. Very soon the inventiveness it inspired had produced something that more closely resembled the familiar desktop computer of today. It was extraordinary how rapidly it happened once the hackers and hobbyists were brought together by

Roberts's minicomputer: another instance of the big technological breakthrough coming not from the established computer companies but from enthusiastic and ingenious amateurs.

In California in the early 1970s a movement arose in opposition to what it regarded as the oppressive control of computers by big business and the military. There was Community Memory which set out to make computer terminals available on the streets, and then the People's Computer Company which published a newsletter and had as its manifesto: 'Computers are mostly used against people instead of for people; used to control people instead of to free them: Time to change all that.' The arrival of the Altair 8800 in this counter-culture almost immediately spawned an ad hoc group of computer hackers and hobbyists who answered a flyer calling for a meeting on 5 March 1975 in the twin garage of a home in Menlo Park owned by an out-of-work engineer Gordon French:

AMATEUR COMPUTER USERS GROUP
HOMEBREW COMPUTER CLUB . . . you name it
Are you up to building your own computer? Terminal? TV Typewriter?
I/O device? Or some other digital black magic box?
Or are you buying time on a time-sharing service?
If so, you might like to come to a gathering of people with like-minded interests. Exchange information, swap ideas, help work on a project, whatever. . .

Thirty-two enthusiasts turned up on the night for that very first meeting of what was to become the hugely important Homebrew Computer Club. Among them was a young man who had a full-time job with Hewlett-Packard designing electronic calculators. A friend had seen the flyer and thought the club was for hobbyists making TV terminals, which is what he was interested in. If he had been told it was about microprocessors, Steve Wozniak recalled in his autobiography *iWoz*, he would not have gone. He was shy, he had not kept up his early interest in computers, and he felt out of his depth:

> It was cold and kind of sprinkling outside, but they left the garage door open and set up chairs inside. So I'm just sitting there, listening to the big discussion going on. They were talking about some microprocessor computer kit being up for sale. And they all seemed excited about it. . . . So it turned out all these people were really Altair enthusiasts, not TV Terminal people like I thought. And they were throwing around words and terms I had never heard . . . chips like the Intel 8080, the Intel 8008, the 4004. . . . I'd been designing calculators for the last three years so I didn't have a clue.

That night Wozniak had a look at the data sheet for the microprocessor and suddenly realised that it was similar to a computer he had designed a few years earlier with a friend. He had called it his Cream Soda computer because that was what they enjoyed when they were making it. The revolutionary element of the Altair was the all-in-one chip, which would do the work of several chips in his original design. Fired with a new enthusiasm, Wozniak felt confident that he did not need an Altair. He could buy the component parts for his computer and there was a microprocessor just on the market made by a company in Pennsylvania, which he bought over the counter at an electronics fair in San Francisco.

By Sunday 29 June 1975 Wozniak had a computer working with a keyboard and a screen. 'It was the first time in history,' he recalled, 'that anyone had typed a character on a keyboard and seen it show up on their own computer's screen right in front of them.' At meetings of the Homebrew club, which were held every other Wednesday, Wozniak would demonstrate his trail-blazing computer. He xeroxed the design and handed it out. He was very proud of it. On one or two occasions he had brought along a friend called Steve Jobs, who helped him carry equipment. Jobs began to think in terms of selling parts that Wozniak designed. He reasoned: 'Well, even if we lose our money, we will have a company. For once in our lives we will have a company.' Jobs suggested they called the company Apple. Wozniak worried that it might upset the Beatles. They went for it anyway and Wozniak's computer became known as Apple 1. It was the Apple 2, which went on sale in 1977, that became the first successful computer for the home, another triumph for the amateur over big business. Once it was clear there was a market for the Apple, the major computer companies – which had been making much larger machines for science laboratories and industry – produced their own version of what became known as the PC.

The enthusiasts of the Homebrew Computer Club were the beneficiaries of several strands of history which, in their early conception, had nothing to do with creating 'thinking machines' at all. They had their origins in the nineteenth century and were not to be brought together until the 1970s. One was an automated form of silk weaving: another – and this is the more surprising – a revolutionary form of printing, which became essential for the creation of microchips. The system of logic that was drawn on to bring about the digital revolution was also devised more than a century before the first microprocessors appeared. And, in each case, just like the Altair enthusiasts of the 1970s, the pioneers included loners and amateurs. The technological histories that led to the development of the modern computer and the shrinking of its component parts so that it could be transformed into a tiny chip can both be traced back to the late eighteenth century.

* * *

In the history of computers, replete with a hundred brilliantly devised, but grindingly unromantic calculating machines, there is one invention that glimmers with colour and charm. It produced not mathematical calculations but beautiful patterns created with intricately woven threads of silk. These included an astonishingly detailed portrait of the inventor himself in his office with a model of his machinery. One of the few of these produced – the portrait could be replicated exactly with a pre-set program – had pride of place in the home of the wealthy and eccentric English mathematician, scientist and inventor Charles Babbage. He liked to show it off to guests and, as they admired it, to ask them to guess how they thought it was created.

In his memoir, *Passages from the Life of a Philosopher*, Babbage tells the story of a visit paid to his London house in 1842 by Queen Victoria's uncle, Count Mensdorf, accompanied by the Duke of Wellington. Very happy to oblige royalty and the duke, Babbage looked forward to showing them his prized portrait of the Frenchman Joseph-Marie Jacquard as well as his own invention of a calculating machine he called the Difference Engine. The time had been fixed for two o'clock in the afternoon when a message came from the duke to say Prince Albert would like to join them, and could they make it an hour earlier.

'I must freely admit,' Babbage recalled,

> that I did not greatly rejoice at this addition to the party. I resolved, however, strictly to perform the duties thus thrown upon me as a host, as well as those to which Prince Albert was entitled by his elevated position. . . . I asked his Royal Highness to allow me to show him a portrait of Jacquard . . . as it would greatly assist in the explaining of the nature of calculating machines.
>
> When we had arrived in front of the portrait, I pointed it out as the object to which I solicited the Prince's attention. 'Oh! That engraving,' remarked the Duke of Wellington. 'No!' said Prince Albert to the Duke. 'It is not an engraving.' I felt for a moment very great surprise; this was changed into a much more agreeable feeling, when the Prince instantly added. 'I have seen it before.'

When he had first seen it, no doubt the Prince would have been deceived as all Babbage's previous guests had been. However, he had already learned that it was not an engraving.

The picture, a copy of an original painting, was of such fine texture that it had fooled some of the most eminent artists of the day. 'A short time after I became possessed of this beautiful work of art,' Babbage wrote, 'I met Wilkie [possibly Sir David, the Scottish painter] and invited him to come and see my recent acquisition. He called on me one morning. I placed him a short distance in front of the portrait, which he admired greatly. I then asked him what he thought it was. He answered "An engraving!" On which I asked "Of what kind?" To this he replied "Line-engraving to be sure!" I drew him a little nearer. He then mentioned another style of engraving. At last, having placed Wilkie closer to the portrait, he said, after a considerable pause: "Can it be lithography?"" This encounter was around 1840, when lithography was all the rage in Europe and had inspired the creation of the first photographs.

The delight Babbage experienced in flummoxing a famous artist was matched by his excitement in explaining to him that the way in which this astonishing textile was produced employed essentially the same programming technology as an instrument he was designing and was in the process of building. Instead of making cloth it would be used to solve instantly all kinds of complex mathematical problems. The key to it was the use of 'punched cards', which in the Jacquard loom controlled the action of hundreds, or even thousands, of rods, which worked threads that were interwoven automatically according to a blueprint. The card would allow a particular thread through if the rod controlling it hit a strategically placed gap, and prevented it from shifting if there was no gap. Effectively, the punched cards turned threads on or off according to their place in the preordained design for the cloth. The portrait of Jacquard was made after his death by silk weavers in his home town of Lyons to demonstrate the extraordinary sophistication of the automated loom he had invented. Twenty-four thousand punched cards were used to achieve the very fine black and white weave that proved so deceptive even to the trained eye of an artist.

Jacquard provided no account of what led him to devise the programmable, punched-card loom, leaving contemporaries and historians to speculate. We know for certain that Joseph-Marie was born on 7 July 1752 in Lyons, then the centre of France's important silk weaving industry. The family surname was Charles; Jacquard was a nickname they adopted. His father Jean-Charles had a reasonably prosperous business making brocade and his wife Antoinette gave birth to nine children, only two of whom, Joseph-Marie and a sister Clemence, survived. Like other children in Lyons, Joseph (as he was usually known) went straight to work on the looms, probably as a draw boy, carrying out the tedious job of pulling threads, which his punched cards would later make redundant.

Joseph had no formal schooling but was taught to read and write when he was twelve by a family friend who married Clemence. Little is known about his adolescence. In 1772 his father died, leaving him with a profitable business and some valuable assets. He appears to have had little interest in the weaving trade and lived largely on his inheritance. In 1778 he married and a year later a son was born. He continued to squander his father's money and by 1783 everything was gone. He sold his last possessions, and, according to legend, he left his wife in Lyons earning a pittance in a factory making straw hats while he became an itinerant labourer. There is no reliable record of where he went or what he did.

Scraps of biography have him defending Lyons against the Revolutionaries in 1793 with his son fighting alongside him. When they were defeated they escaped the guillotine, probably changed their names, and became turncoats, joining the Revolutionary Army. At some point his fifteen-year-old son was killed and Joseph returned to Lyons in 1798. He was wounded and spent some time in hospital before taking on odd jobs. It was only then that he embarked on an astonishing few years of creativity and invention.

Jacquard's first modification of the traditional loom was simply the creation of a treadle to replace the 'draw boy'. He took out a patent in 1800 and exhibited his new loom in Paris in 1801 where it won a bronze medal. The following year he won a substantial prize for designing an improved loom for weaving fishing nets in a competition sponsored by the French Society for the Furtherance of National Industry. Then in 1804 Jacquard patented his loom, using punched cards to program the actions of the warp thread. In fact, the punched card was not his innovation: it had been tried without much success by French weavers before. But Jacquard had the skill to make it practical and efficient. Somehow he had learned to become an accomplished craftsman.

As James Essinger points out in his book, *Jacquard's Web: How a Hand-loom Led to the Birth of the Information Age*, this was an auspicious time to be a talented inventor of machinery in France. Napoleon was keen to promote new technology and in particular any that might steal a march on British rivals who dominated the textile industry. Jacquard attracted investors and the admiration of Napoleon himself who, with his wife, Josephine, visited the Lyons workshop in 1805. Shortly afterwards the loom was declared to be public property in France and Jacquard was compensated with a handsome pension for life.

The loom was a truly astounding achievement and deserved the honours that were showered on Jacquard towards the end of his life. It was estimated that once the punched cards had been programmed it would weave twenty-four times faster than other looms and the same pattern could be repeated indefinitely. A new arrangement of the punched cards could be applied to the same loom to create a different pattern. Copies of paintings or etchings were made by pixelation, the arrangement of dots. So superior was the loom for silk weaving that it was soon copied in Britain, and later in America. The Smithsonian has a collection of portraits of American presidents created on a later Jacquard loom. Originally worked by hand, it was later powered by steam and won the admiration of all who saw it in operation. In a book on invention entitled *Knowledge is Power* published in 1859, Charles Knight said of the Jacquard loom at work in England:

> Those who would comprehend the extent of ingenuity involved in the principle of this invention, and the beautiful results of which it is capable, should witness its operation in a Halifax power-loom. In a bobbin-net machine the cards are connected with a revolving pentagonal bar, each side of which is pierced with holes, corresponding with the pins or levers above.
>
> When a card comes over the topmost side of the pentagon the levers drop; but those pins only which enter through the holes in the card affect the pattern which is being worked. Any one who views this complicated arrangement in a Nottingham lace-machine, requires no small amount of attention to comprehend its mysterious movements; and when the connection is perceived between that

> chain of dropping cards, and the flower that is being worked in the lace, a vague sense of the manifold power of invention comes over the mind – we had almost said an awful sense.

The mechanism of the Jacquard loom fascinated Babbage and encouraged him to pursue his dream of creating what he called an Analytical Engine, a prototype mechanical computer. It was an endeavour which proved to be over-ambitious and doomed to failure despite the fact that for a time he had the most fervent and loyal advocate of his work in Ada, Countess of Lovelace.

* * *

Born in 1791, Charles Babbage was the son of Benjamin Babbage, a wealthy London banker, and Elizabeth Teape. Both parents came from Totnes in Devon and belonged to upper-class families. When he was ten Charles was sent to Devon to recover after suffering a fever and from that time on his education was disrupted. Early on, however, he developed an aptitude for mathematics and went to Cambridge when he was nineteen where he enjoyed a rich social life. During the holidays he visited Devon, where he fell in love with Georgiana Whitmore and was determined to marry her despite his father's objection that Charles had no position or income and was only twenty-three. In defiance, he took with him from London a tutor, a clergyman who agreed to perform the wedding ceremony in Devon. Despite his objections to the marriage, his father gave Charles an allowance and with rents from property he could live reasonably well. He and Georgiana settled in London in 1815.

In the years before he devoted himself to his calculating engines he had become a controversial figure in the scientific world of London, disputing with the Royal Society and holding a number of senior positions, but without anything in the way of paid employment. He was aware that his father thought him a failure, while his mother doted on him. In 1822 he produced a model of a Difference Engine to demonstrate his concept of a calculating machine and began to work on a full-scale version. He was a married man with a growing family and might have hoped he would at last appear to his father as a success. Then tragedy struck not long after he and Georgiana had returned from a successful visit to Paris, where he had met many of the leading French scientists.

In one grim year, 1827, Georgiana and two of her sons died. Of her eight children two others had died in infancy; only four survived to adulthood. Babbage lost not only his wife and two children but also his father, from whom he inherited a small fortune. Taking with him a travelling companion, Babbage left his surviving children to be cared for and set off on a tour of Europe, which lasted just over a year. He did not remarry. His work on the calculating machines became more engrossing and absorbed more and more of his considerable income. He ended his life a lonely widower surrounded by the unassembled bits and pieces of his Analytical Engine.

A problem with the legacy of Babbage's inventions is that, despite their tremendous ingenuity and even genius they were of absolutely no practical use. His first great project, which became known as Difference Engine No. 1, cost the British Government more than £17,000 before the Prime Minister Sir Robert Peel told Babbage in person that he would get no more funding. Their brief meeting in November 1842 ended with Babbage – irascible as always – storming out. He had anyway, by then, embarked on a more sophisticated invention, his Analytical Engine. This was to be an automatic calculating machine, which would be 'programmed' to carry out various mathematical calculations with the punched-card system that Jacquard had perfected for silk weaving.

Both these calculating machines were necessarily entirely mechanical, as Babbage did not have available anything electronic. There are a number of stories about the genesis of the idea that routine mathematical tables might be compiled by a machine rather than tediously by computers – who were at that time, and for long after, people. Babbage was an admirer of Napoleon and an ardent Francophile who regarded continental science as far superior to that of science in Cambridge. He was impressed by the way an aristocrat, Baron Gaspard de Prony, who had narrowly escaped the guillotine, set about the task after the Revolution of providing the French Ordnance Survey with tables to help in the calculation of property values. De Prony decided to adopt the 'division of labour' principle that had been expounded by Adam Smith in his *Wealth of Nations* published in 1776. Applied to mathematical calculation this meant appointing a handful of the very best brains to oversee the project, a less skilled team to break it down into manageable calculations, and a third team, by far the largest, to perform tens of thousands of relatively simple additions. A large contingent of this computing workforce was drawn from the legions of hairdressers rendered redundant by the rolling of aristocratic heads during the Revolution: these computers were, in a sense, early nineteenth-century equivalents of the transistor.

Babbage, interested in industry and mechanics, was very conscious that the world was being transformed by steam power and tireless machines that were replacing the exertions of men and horses. In his *Passages* he tells the following story:

> One evening I was sitting in the rooms of the Analytical Society, at Cambridge, my head leaning forward on the Table in a kind of dreamy mood, with a Table of logarithms lying open before me. Another member, coming into the room, and seeing me half asleep, called out: 'Well Babbage, what are you dreaming about?' To which I replied, 'I am thinking that all these Tables (pointing to the logarithms) might be calculated by machinery.'

Another anecdote has him saying to his close friend John Herschel that he wished the astronomical tables could be created by 'steam'. They had been correcting tables and found inaccuracies that were difficult to put right. A machine would be unerring and would not fall asleep through boredom.

The Analytical Engine unashamedly borrowed as a model the Jacquard loom, which Babbage said was a 'nearly perfect' analogy. There would be two sets of punched cards, 'those to direct the nature of the operations to be performed . . . the other to direct the particular variables on which those cards are required to operate'.

What was difficult for his contemporaries to understand was that his machines worked automatically: a mass of interconnecting cogs replicated in a sense the working of the human brain, or, at least, that part of it which was used to solve mathematical problems. One of those shown the Difference Engine, Lady Byron, referred to it as Babbage's 'intelligence machine'. This was in 1832, and the estranged wife of the aristocratic poet Lord Byron had taken to the soirée in Dorset Street, London her daughter Ada who was then seventeen and had been brought up to take an interest in mathematics. Lady Byron was a mathematician herself and she wanted, at all costs, to keep her daughter away from the poetic sensibility of her father. However Ada, entranced by the intelligence machine, wrote about it with a poetic charm that would have been quite impossible for the inventor himself.

The only legitimate daughter of Lord Byron, Ada was married at the age of twenty to the aristocrat William King who, in 1838, was created Earl of Lovelace. She was mistress of three substantial homes and gave birth to three children, and yet found the time to study Babbage's machines and to make them intelligible to a wider audience. She translated from the French a book about them written by an Italian admirer of Babbage and added her own extensive notes. The work, published in *Scientific Memoirs*, vol. 3 in 1842, included the vivid observation that the Analytic Engine 'weaves algebraic patterns just as the Jacquard loom weaves flowers and leaves'. Lady Lovelace imagined the Analytical Engine might compose 'elaborate and scientific pieces of music of any degree of complexity and extent'.

In contrast to her intellectual vigour, Ada Lovelace's health was never robust. Her writing was affected by the laudanum prescribed by her doctors, and she moved into a social set where gambling on horses was popular. When she became seriously ill her husband called on her mother, Lady Byron, for help. The gambling was stopped and Lady Byron took over the household. She refused to allow Ada to see Babbage.

Ada died on 27 November 1852 at the age of just thirty-six. She had suffered an ignominious end, but she was not forgotten. There is a plaque in St James's Square a few doors down from the London Library which commemorates her as a pioneer of the computer and in 1979 the official computer software language of the US Department of Defense was given the name Ada in her honour.

Babbage's youngest son Henry described his father's sad and surreal death at 1 Dorset Street in his *Memoirs and Correspondence of Major General H.P. Babbage*:

> On the 16th the organ-grinders were particularly troublesome, and before I went to sleep I wrote again to the Commissioner of Police, but the organs were playing all the same on the 17th, both about midday and about 9p.m. and in the afternoons there was a man inciting boys to make a row with an old tin pail . . . no policeman in sight. At 8.45p.m. there was again more organ playing. On [the] 18th organs playing again about midday. . . . C.B. passed away about thirty-five minutes past 11 o'clock p.m. Shortly before he had said 'What is the clock Henry?' and I told him.

For years Babbage had waged a war against street musicians who drove him mad with rage. In his memoir he included a whole chapter on 'street nuisances', listing the *Instruments of torture permitted by the Government to be in daily and nightly use in the streets of London:* organs, brass bands, fiddles, harps, harpsichords, hurdy-gurdies, flageolets, drums, bagpipes, accordions, halfpenny whistles, Tom-toms, trumpets and all kinds of singing and shouting. This cacophony, he complained, 'destroys the time and energies of all the intellectual classes of society by its continual interruptions of their pursuits'. In his own case he thought the nuisance of Italian organ grinders, German brass bands and Indian Tom-Tom drummers had sapped a quarter of his mental energy.

Street musicians made a terrible noise outside smart houses, not to be rewarded for their entertainment but to bully the owner into paying them to take their din elsewhere. Babbage joined a campaign to have the law changed and in doing so made his home address a target. On occasion the racket was sustained for five hours and the 'instruments of torture' were there to the bitter end. Few attended his burial at Kensal Green cemetery: there was just one 'carriage' for the man who had once held some of the most brilliant soirées in London. The inscription on his tombstone simply reads:

> Charles Babbage, Esq. Born 26 December 1791, Died 18 October 1871.

No mention of calculating machines: no epitaph at all.

Although Babbage did not finish building any of his machines, a Swedish printer, publisher and journalist Georg Scheutz read a glowing and largely inaccurate account of the Difference Engine and set about making his own version of it. His fifteen-year-old son, Edvard, built a prototype with no knowledge of Babbage's machinery. This was during Babbage's lifetime and Scheutz was anxious when they exhibited their machine at the Royal Society in London in 1854: would Babbage accuse them of plagiarism? In fact he championed them in their efforts to promote and raise funds for building a more sophisticated machine. Their calculating engine was much admired and was viewed by all the leading scientists of the day, including Michael Faraday and Charles Wheatstone. It was shown around the world and won a gold medal in a major exhibition in Paris in 1855. One machine was even used by the Registrar General in London. But all in the end failed and both father and son died bankrupt.

Just as powered flight was a hopeless dream in the age of the steam engine so was Babbage's dream of the mechanical computer. When the vacuum tube brought electronics to computing in the 1930s the brilliant English mathematician Alan Turing advanced the science of computing with his 1936 paper on computable numbers. But these first electronic computers were huge and built for specific purposes. They were notoriously noisy and the thousands of valves that powered them generated a great deal of heat. And their use was severely limited by what became known as 'the tyranny of numbers': to be more powerful they had to get bigger and bigger. Valuable in wartime when used for decrypting coded messages, after the war the business world had no use for them and relied for analysis of information on tabulating machines fed with punched cards just like Jacquard's looms. Employing the inventions of the American Herman Hollerith, which led to the formation in 1924 of IBM (International Business Machines), the punched card remained the dominant analytical tool after 1945. Computers were too cumbersome and costly until a major step towards miniaturization was made at Bell Laboratories where the new discipline of solid state physics promised a replacement for the vacuum tube. It was a response to a demand not from the manufacturers of computers but from the overloaded telephone exchanges in America.

* * *

An innocuous and rather dull photograph showing three research scientists wearing shirts and ties and focusing on some miniature component has for years accompanied stories about the invention of the transistor. It is obviously posed and has the hallmark of a corporate promotional shot. It was the cover illustration for the September 1948 edition of the American technical magazine *Electronics* and captioned 'Revolutionary Amplifier, the CRYSTAL TRIODE'. All three of these scientists working for Bell Laboratories, the research arm of the giant AT&T sometimes nicknamed the Ideas Factory, were later to be awarded the Nobel Prize for Physics. They were the first to make a working alternative to the valve or vacuum tube which had revolutionised wireless communication in the early years of the twentieth century (see chapters Two and Five). However, the story behind the photograph is not one of harmony in the research laboratory but one of deviousness and some bitterness.

The man at the workbench, apparently making the breakthrough, is William Shockley and the two men looking over his shoulder Walter Brattain and John Bardeen. According to the authors of *Crystal Fire: The Birth of the Information Age,* Bardeen remarked years later: 'Boy,

Walter sure hates this picture'. Brattain's anger is understandable for it was he and Bardeen who had made the breakthrough and produced the first crude working transistor a year earlier while Shockley was otherwise engaged. There was huge excitement in the lab as well as great secrecy about the invention until finally its practicality was demonstrated to senior management on Christmas Eve 1947 when the voices of Bardeen and Brattain were amplified with transistors rather than vacuum tubes and played through earphones. Straight away Bell Labs' patent lawyers were set to work. Employees of Bell could not claim a patent for themselves as they had to sign away their rights for a nominal $1 when they took the job. But names had to be attached to patents and the lawyers wanted to know whom to credit. Bardeen and Brattain said it was a joint effort by just the two of them.

However, Shockley was their boss and had been pursuing a way of creating a semiconductor device for a long time. He could not abide the thought that his name would not be on the patent. Brattain recalled: 'He called Bardeen and I in separately, shortly after the demonstration, and told us that sometimes the people who do the work don't get the credit for it . . . He thought then that he could write a patent . . . on the whole damn thing. . . . I told him "Oh hell, Shockley, there's enough glory in this for everyone."' The patent lawyers would not accept Shockley's claim to have a part in the invention and, to his fury, he was left in the publicity photograph but off the patent application.

In fact that first transistor was an adaptation of a form of semiconductor which had been widely used in the early days of radio, the so-called 'cat's whisker'. The first to design a radio receiver with a thin wire touching the surface of a crystal was probably the Indian scientist Jagadish Chandra Bose, who demonstrated it in 1894 and took out a patent in 1901. It was a hit and miss device: the wire had to touch the crystal, which might be galena or germanium, at the right point for it to work. The great advantage of the crystal set over the valve wireless was that it needed no power: the signal was enough to receive Morse Code or speech in headphones. In time the crystals were enclosed in capsules and became much more reliable. But they could not amplify a signal or any other electric current as a vacuum tube could and their use was therefore limited.

The raw material for a transistor had been around for a very long time and so too had the knowledge that certain minerals had semiconductor properties under certain conditions: they allowed a current to run one way and not the other and their conductivity could be varied. Germanium, used by Brattain and Bardeen for their first transistor, had been isolated and identified in the 1880s. Silicon had been discovered much earlier, finally identified in 1827 by the diligent analysis of Jacob Berzelius (see Chapter Two). Neither element occurs in a pure state naturally and both can take many different forms. However, making use of them for electronic circuits required an understanding of their properties that came with the revolutionary theory of quantum mechanics developed in the early twentieth century. Shockley, Bardeen and Brattain were among the first generation of physicists brought up on the mind-boggling complexities of the new science.

Of the three, though he usurped the title of co-inventor of the transistor, Shockley was by reputation the most brilliant theoretical physicist. But he was also a troubled man: after his death a suicide note he had left for his first wife, and an account of his attempt to blow his brains out with a pistol, were found in a sealed box. The gun failed to fire and he did not pull the trigger a second time. He was an accomplished rock climber and at one time threatened to head out to do his climbing 'at midnight and unroped'. It is clear he was a dreadful manager, a side of his character which, by a strange quirk of fate, perhaps hastened the development of the microprocessor.

Shockley's father, also William, kept an intimate diary in which he noted that his son, from a young age, had a terrible temper. Young William's upbringing perhaps did not help to improve his social skills, he was an only child and had little chance to meet other children in his early years. His mother, May, was a very colourful figure with something of the Wild West about her as well as a brilliant mathematical brain. Brought up in New Mexico and Missouri by her mother and stepfather, she rode the plains on horseback shooting or trapping rabbits. She studied mathematics at Stanford University and took an interest in geology and rock climbing. Her stepfather was a surveyor and she followed him to Tonopah, Nevada to work with him. Here she became the first ever woman US Deputy Mineral Surveyor. It was here, too, when she was thirty years old, that she met William Shockley, a mining engineer, who was twenty years older. He was a cultured man who spoke several languages and impressed May enough to persuade her to marry him. In 1909 as newly-weds they set sail for London, where their son William was born the following year.

The family moved back to America after two years and continued a peripatetic life before settling in Palo Alto, California. Young Bill was educated at home much of the time and learned a great deal about engineering and science from his father. For a while he wore the uniform of the Palo Alto Military Academy where, to his parents' great surprise, he received high praise for his conduct and was made an Honor Cadet.

He won a place at the University of California and then, in 1928, moved to the California Institute of Technology, always known as Caltech, where he was taught by Nobel Prize winners and was immersed in the exciting world of quantum mechanics and Einstein's physics. After he graduated from Caltech he took a doctorate at the Massachusetts Institute of Technology and went on to work for Bell Laboratories.

When Shockley arrived at Bell in 1937 Walter Brattain had been working there for eight years and had a reputation for being one of the best experimental physicists. Brattain had been born in China to American parents but was brought up from infancy in Washington State. When he was still a boy his father bought a ranch and he and his brother became cowboys, herding livestock in the hills and shooting rattlesnakes. He studied at Whitman College, following in the footsteps of his parents. Later he recalled that the generation before him had gone to Germany to keep in the forefront of the new physics: his was the first that could afford to stay in America, which was catching up.

Brattain and Shockley became friends at Bell Labs and with their wives and children would sometimes camp together on an island in Lake George in upstate New York to which the Shockleys made regular trips. Although he did not join Bell Labs until 1945, John Bardeen knew both Brattain and Shockley before the war. He knew Brattain's brother Robert when they were both at Princeton, and Shockley when he was a student at Harvard. Bardeen, who came from an academic family, studied at the University of Wisconsin where his father was Dean of the Medical School. He began his career in the oil industry but left it to return to college, taking a course in physics at Princeton. All three abandoned their studies and research when the Japanese attack on Pearl Harbor brought the United States into the Second World War. Shockley was given a high-level and extremely dangerous assignment, crossing the Atlantic a number of times to discuss with the British ways to combat the German U-boats.

When the search for an alternative to the vacuum tube began in earnest at the end of the war the interest in AT&T was not in computers but in the elaborate switching systems of the telephone exchanges. In fact there is no mention of computers in the reminiscences of these three scientists. Their field was solid state physics and the laboratory exploration of the behaviour of elements. Theory and experiment went hand in hand, one complementing the other. It was trial and error that finally got the first transistor to work. Shockley, furious to have missed out on it, soon dreamed up a better version. But transistors were still little bug-like things with wires sticking out of them.

Despite his scientific brilliance, Shockley did not get the promotion he thought he deserved. The reason for this was undoubtedly his 'management style'. He also had a habit of playing practical jokes which were not always regarded as in the best possible taste: a mechanical duck, for example, disrupting a lecture by a distinguished visiting scientist. He was a prankster and an accomplished magician who would perform at children's parties. It was perhaps inevitable that Shockley's individualism would lead to his decision to set up his own company to specialise in the manufacture of transistors, which he anticipated would one day replace valves in all kinds of electronic devices. In 1955, when he was forty-five years old, Shockley divorced his first wife Jean, left Bell Labs and moved to Palo Alto. At this time Shockley was sure that the key element in the creation of new solid state electrical components was silicon and it is sometimes said that it was he who 'put the silicon into Silicon Valley'.

There is no doubt that the silicon transistor ushered in an entirely new electronic era and it promised to solve the dilemma of the computer's 'tyranny of numbers'. But there was another, even more momentous, innovation emerging from the Ideas Factory of Bell Labs. A brilliant mathematician had applied a system of logic that had been developed by George Boole, a contemporary of Charles Babbage, to the problem of switching in telephone exchanges. Boole had taken only a passing interest in Babbage's mechanical computers but he had, by chance, provided the logic which ushered in the digital age. Boolean is now a term familiar around the world but very few know what a remarkable man he was.

* * *

It is not often that the trade of shoemaking is given any prominence in the history of the computer, but, as an influence, it has as much claim in its way as silk weaving. In the case of the cobbler it was his proverbial fascination with books and learning rather than mechanisation that was significant. For reasons that have never been satisfactorily explained, shoemakers across many cultures were by reputation amongst the most learned of artisans. Born in 1777 at Broxholme in Lincolnshire into a farming and trading family, John Boole proved to be true to type. Apprenticed to a shoemaker in Lincolnshire at the age of fourteen he served his seven years and hated every minute of it. However, that was his trade and when he was twenty-three he moved to London, where he found conditions no better. Here, as his biographer put it, he took 'consolation in learning'. He kept a French dictionary in a drawer with his tools and took an interest in mathematics and the making of telescopes and microscopes.

While in London, Boole met the lady's maid Mary Ann Joyce who became his wife. They moved back to Lincoln where he opened his own boot and shoe shop. Though they were married in 1806 it was not until

1815 that Mary gave birth to their first child, George, and then three other children in quick succession. John Boole was not a very successful shoemaker but he was a quite exceptional self-taught scientist and mechanic. He made his own telescope and put in his shop window a little notice: 'Anyone who wishes to observe the works of God in a spirit of reverence is invited to come and look through my telescope.' John spent more time with the Lincoln Mechanics' Institute, of which he was a founder member, and campaigning for half-day closing, than he did in his shop. There was little money for George's education, which began when he was just eighteen months old. He went from one Dame school to another before attending a small commercial school run by a friend of his father.

He was by all accounts a prodigy. The tale was told that he went missing one day still dressed, as young boys were, in frocks and pinafore and was found in a busy street in Lincoln surrounded by an excited crowd, which was throwing him coins for spelling difficult words correctly. He learned a great deal from his father, who noted that at the age of eleven George was reading advanced books on geometry and devoured everything he could on mathematics. He learned Latin, first from his father and then from a bookseller in Lincoln who lent him books from his large library. To Latin he added Greek and at the age of fourteen translated Horace's 'Ode to Spring' into English and had it published in the *Lincoln Herald*. Not only did he learn at an astonishing rate, he had a prodigious photographic memory, of which he wrote:

> This does not result so much from strength of memory as from the power of arrangement which provides its proper place in the mind for every fact and idea and thus enables me to find at once what I want just as you would know in a well-ordered set of drawers where to lay your hand in a moment upon any article you require.

As his biographer Desmond MacHale remarked: 'One could ask for no better intuitive description of information storage and retrieval!'

Boole had contemplated a career as a clergyman but he had his doubts and instead took a job teaching at a boarding school in Doncaster, a town about forty miles from Lincoln. He hated the place: he felt isolated and frustrated that he could not get on with his mathematical studies as he would have liked. However, it was here that he had what turned out to be his revelation. According to MacHale:

> the thought flashed upon him suddenly one afternoon as he was walking across a field, but he laid it aside for many years. . . . The thought however smouldered in his subconscious and became an integral part of his main ambition in life – to explain the logic of human thought and to delve analytically into the spiritual aspects of man's nature.

After two years in Doncaster Boole took another teaching post, this time in Liverpool. But he did not stay long. At the age of twenty he decided to open his own school for boys and girls in Lincoln. He made a success of it and founded more schools, one of which also provided a home for his family. At the same time he began to contribute papers on mathematics to various journals and to gain a reputation as an original thinker. For fourteen years after the founding of his first school he made a living as a schoolmaster and used what free time he had to develop his mathematical ideas. He did not marry and was still a single man when he applied for a professorship at the newly founded Queen's College, Cork in Ireland. The last two lines of his letter of application read: 'I am able to read scientific papers in the French, German and Italian languages. I am not a member of any University and have never studied in a college.'

The application was made in 1846 and was accompanied by testimonials from a number of leading mathematicians, three of them from Cambridge University, as well as a touching recommendation from the mayor and some of the leading citizens of Lincoln. Boole waited for three years before he heard that he had been given the professorship of mathematics. In the interval his father died and he, at the age of thirty-three, became the main breadwinner for the family. At the news of his appointment a grand dinner was held at the White Hart in Lincoln at which he was presented with a silver inkstand and some finely bound books bought by public subscription.

Shortly after George had become settled in Cork he met an eighteen-year-old girl, Mary Everest, the niece of John Ryall, the Professor of Greek. Another uncle was Colonel Sir George Everest (pronounced eevrest) after whom the Himalayan mountain was named. Mary was intrigued by Boole who, she noted, was very popular at social gatherings and considered absolutely trustworthy. Her education had been disrupted when her father took the family to France in search of a cure for a lingering illness and, when they returned, George was asked to give her some lessons. Mary was interested in mathematics and she began to exchange letters with him on the subject. In 1852 he visited her family at their home in Wickwar in Gloucestershire where Mary's father was the rector. For three years they corresponded, mostly about mathematics, at a time, 1854, when Boole was writing his seminal work *An Investigation of the Laws of Thought on Which are Founded the Mathematical Theories of Logic and Probabilities*. There does not appear to have been any romantic attachment and it came as a great surprise to Mary's family and friends when George proposed and she accepted. It was the death of her father in 1855 that appears to have prompted George to offer Mary the protection of marriage.

They were well suited and Mary took a great interest in George's writing and thought. She was an admirer of Babbage who, she said, had known her father, and wrote of him in a book *The Message of Psychic Science*:

> if I were asked to point out the two greatest benefactors to humanity in this century, I think I should be inclined to mention Mr Babbage, who made a machine for working out series and Mr Jevons who made a machine for stringing together syllogisms. Between them they proved, by unanswerable logic of facts that calculation and reasoning, like weaving and ploughing are work, not for human souls, but for clever combinations of iron and wood.

Although he had little contact with Babbage, Boole did visit him when he was in London and was shown the Difference Engine; 'it was a pleasure and an honour to meet you,' he wrote.

In the nine years between their marriage and George's early death from pneumonia in 1864 he and Mary had five daughters, all of whom had remarkable careers. Mary continued to promote her late husband's work, and to survive took on a variety of secretarial jobs. As she aged she became increasingly eccentric and died in 1916.

It was George Boole, who showed little interest in calculating machines yet tried to fathom the way in which people think and reason, who provided the inspiration for not only the modern computer but the entire world of digital electronics. The mathematics historian Ivor Grattan-Guinness says in his entry for Boole in the *Oxford Dictionary of National Biography* that he would have 'hated the computer'. He argues that Boole would not have liked its 'repetitiveness'. Boole's biographer Desmond MacHale disagrees, believing that Boole would be 'immensely proud' of the contribution of his logical system to the development of the computer and the communications revolution.

It was a young American who realised that Boolean algebra could be used to process information electronically. Claude Shannon's thesis, written in 1937, had the unexciting title 'A Symbolic Analysis of Relay and Switching Circuits' but came to be regarded as the most influential paper of twentieth-century electronics. As with so much innovation at the time, it was a proposal for solving some serious problems with the telephone networks, which were becoming overloaded. Working for Bell Labs, Shannon used Boolean logic to devise a way of sending information in the form of pulses rather than waves. His revelation had come about because he had, most unusually, taken courses in both logic and electronics.

In his history of the microprocessor, *The Chip: How Two Americans Invented the Microchip and Launched a Revolution*, T.R. Reid says of Shannon: 'If society allocated fame and fortune on the basis of intellectual merit, Claude Shannon would have been as rich and famous as any rock idol or football star.' Shannon was brought up in the small town of Gaylord, Michigan in 1916; his father was a judge and businessman and his mother the principal of the high school. He studied electrical engineering and mathematics at the University of Michigan, graduating in 1936. From there he went to MIT, where he had the opportunity to work with distinguished engineer Vannevar Bush on a computer rather like a semi-electronic version of Babbage's Differential Engine. Shannon became intrigued by the work of the relays and wrote his celebrated thesis in which he drew on the logic of George Boole. This work alone, which he completed when he was twenty-two, would have been sufficient to gain him a prominent place in the history of electronics for it showed how Boolean symbolic logic could be used to analyse complex systems such as the switching systems of a telephone exchange.

Later, working at Bell Labs, he went on to propose a theory of communication in which all electronic information could be reduced to a common unit represented as a 1 or 0, what he called a 'binary digit' soon shortened to 'bit'. This became the measure of a computer's memory; more bits, more memory. Though he did little to popularise his work, which is perhaps why he is not well known to the public (like a number of other prominent engineers he has no entry in the *American National Biography*), Shannon is regarded as 'the father of the digital age'. He died aged eighty-four in 2001 after suffering from Alzheimer's for a number of years. His obituary in the London *Times* captured something of his eccentric character under the heading 'Playful genius who invented the bit, separated the medium from the message and laid the foundations for all digital communications'.

> To colleagues in the corridors at the Massachusetts Institute of Technology who used to warn each other about the unsteady advance of Shannon on his unicycle, it may have seemed improbable that he could remain serious for long enough to do any important work. Yet the unicycle was characteristic of his quirky thought processes, and became a topsy-turvy symbol of unorthodox progress towards unexpected theoretical insights...

> Like Charles Babbage, Shannon was known by his contemporaries as 'the Irascible Genius'. When he returned to MIT in 1958, he continued to threaten corridor-walkers on his unicycle, sometimes augmenting the hazard by juggling. No one was ever sure whether these activities were part of some new breakthrough or whether he just found them amusing. He worked, for example, on a motorised pogo-stick, which he claimed would mean he could abandon the unicycle so feared by his colleagues. . .

It was Shannon's revolutionary information theory that provided the logic for the digital age. But the miniaturisation of electronics, which finally made the personal computer a possibility, had a very different and quite independent history. Transistors had enabled a reduction in the size of computers but until the 1970s they were still bug-like components that had to be wired together. The last, and momentous, breakthrough was the creation of the microchip. Today in Silicon Valley these are manufactured in space age factories from which the tiniest speck of dust has been removed. Technicians are dressed like astronauts as they impress on purified silicon disks layers of minuscule components using a technique known as photolithography. Its origins can be traced back to the inspiration of an impoverished German playwright and actor working in the early nineteenth century.

* * *

In his memoir, *The Invention of Lithography*, published in 1818, Alois Senefelder wrote: 'If I had possessed the necessary money, I would have bought types, a press and paper and printing on stone probably would not have been invented so soon. The lack of funds, however, forced me to other expedients.' Senefelder was born in 1771 the son of an actor of Bavarian descent who performed for the court in Prague. His father did not want Alois to follow him on to the stage and insisted that he study law. However, his father died while Alois was still a student and he abandoned his legal studies to fulfil his first ambition – to make his name as an actor and playwright. When he was turned down by the court he became a strolling player, but without success. He did better as a playwright but found the cost of reproducing his scripts prohibitive and sought a way of printing them himself. If he became a printer, he reasoned, he would have a steady income and a means of reproducing his own work. But he had no money.

As he could not afford to buy all the necessary type and presses for printing he decided to teach himself etching on copper plate. This involved learning to write backwards so that the impressed image would be the correct way round. It was tedious and too expensive. As a way of practising he began to work on porous Bavarian limestone. As he recalled in *The Invention of Lithography*:

> I had just ground a stone plate smooth in order to treat it with etching fluid and to pursue on it my practice in reverse writing, when my mother asked me to write a laundry list for her. The laundress was waiting, but we could find no paper. My own supply had been used up by pulling proofs. Even the writing-ink was dried up. Without bothering to look for writing materials, I wrote the list hastily on the clean stone, with my prepared stone ink of wax, soap, and lampblack, intending to copy it as soon as paper was supplied.

According to Senefelder, this impromptu laundry list proved to be the happy chance which led him to the discovery that he need not cut into the stone at all but could create a printing block from which he could make copies using an entirely chemical process based on the fact that water and oil do not mix. It took him a long time to find the right formula for what he called 'chemical printing' and even longer to create a printing press that was reliable. To do this he went into partnership with music publishers. He had to endure a family feud over the rights to his invention and many setbacks before the brilliance of this new and cheap form of printing made him celebrated in Bavaria and known throughout Europe. It was in France that Senefelder's discovery was given the name *lithographie*, meaning 'writing on stone'.

In *Breaking the Mould: The First Hundred Years of Lithography* Michael Twyman makes the point that this was 'the first essentially new method of printing to have been developed since the fifteenth century – the first alternative that is to the well tried methods of printing from type and wood blocks on the one hand or copper plates on the other'. During the nineteenth century lithography dominated printmaking in Europe and North America and many artists made use of it to reproduce their work and to create new art forms. While the principle of 'chemical printing' remained the same, the technology evolved so that metal could be used as well as stone, which meant that lithographic techniques could be used in rotary presses. And Senefelder's discovery did more than revolutionise the art of printing: it inspired the creation of an entirely new way of copying images which in its early days went by the name of heliography.

* * *

On 15 April 1952 *The Times* ran the story of a remarkable find:

> A photograph that establishes Nicéphore Niépce as the first man to obtain a picture from nature with a camera has come to light recently. The photograph is believed to have been taken in 1826, but all trace of it was lost after 1898. It has now come into the possession of Mr and Mrs Helmut Gernsheim, of London, who have a collection of photographs.

The barely discernible image showed a view from Niépce's study window at the family estate of Le Gras in Chalon-sur-Saône. Discovery of the print, which had been locked away in a trunk since 1917, unearthed a fascinating period in which the craze for lithography stimulated experimentation with recording images photographically.

Born in 1765, the son of a prominent lawyer, Niépce studied sciences at a Catholic oratory in Angers. Baptised Joseph, he emerged from the college with the adopted name of Nicéphore after St Nicephorus. He was a young man in the turbulent and violent years of the French Revolution and after college joined the Napoleonic armies fighting in Sardinia and Italy. He left the Army in 1794 and spent some time in Nice in an administrative post before returning to the family estate in Chalon-sur-Saône in 1801.

Nicéphore and his brother Claude were keen amateur scientists and inventors. They hoped to make their fortune by their ingenuity and devised a working internal combustion engine which was powerful enough to propel a boat. They were granted a ten-year patent for what they called their pyrolophore, which ran on hot air. In 1817, with just one year before the patent expired, Claude went to England in an attempt to sell it, settling in Kew on the Thames. Back at Le Gras, Nicéphore began to experiment with fixing images projected on the back of a camera obscura. His inspiration was the craze for lithography that had swept through Europe.

To be a lithographer you had to have some skill at drawing. Niépce's son Isidore could draw but when he left home his father had to devise some new way of creating the images to go on the lithographic stone. What he discovered was something he called 'heliography', that is, 'sun drawing'. Artists had routinely used the camera obscura to project the image or a scene of some kind on to a plate where they could trace it and prepare it for reproduction. Niépce did not have a steady enough hand so he sought some other way of fixing the image. With a light-sensitive solution of his own making he found a way of projecting existing etchings on to paper so that they could be turned into a lithograph. In turn this gave him the idea of making a print from the image projected in a camera obscura.

He set his primitive camera up in his study, and aimed it out of the window at the rooftops of estate buildings. A pewter plate brushed with bitumen of Judea, a light-sensitive substance, was exposed for eight hours. Where the light fell, the bitumen was hardened and this produced the ghostly image which, according to your definition, was, or was not, the first 'true photograph'. The exact date is not known but it is believed it was taken even earlier than was first thought, some time in 1822.

Niépce certainly believed he had created something new and exciting. When he came to London in 1827 to visit his brother Claude, who was in trouble, broke and suffering from mental illness, he brought his heliograph of the view from his study window with him, along with some other images he had created. He showed this to a botanist, Francis Bauer, who was impressed and suggested Niépce show them to the Royal Society. Niépce wrote a memoir, *Notice sur l'héliographie* dated Kew, 8 December 1827, and duly presented it along with specimens he called *Les premiers résultats obtenus spontanément par l'action de la lumière*. The Royal Society wanted more information about the invention and when this was not forthcoming they sent everything back. Niépce tried George IV and the Royal Society of Arts but there was no interest. In the end he gave everything to Francis Bauer and returned to France.

While Niépce failed to excite scientists in London with his invention, he was pursued in France by an artist called Louis Daguerre, who had made his name and a small fortune with his 'dioramas', huge paintings animated with lighting devices which were a popular forerunner of cinema. By chance Daguerre learned of Niépce's researches from an optician who provided him with equipment and he wrote to him to express his interest in his heliography. Lacking funds, Niépce agreed to go into partnership with Daguerre who was then able to learn everything Niépce had discovered. Daguerre improved on the technique of 'sun drawing' but the collaboration ended when Niépce died suddenly. Daguerre promptly named the heliograph the *Daguerrotype* and effectively stole Niépce's thunder. By the 1840s Europe was in the grip of 'Daguerre mania', a sequel to the earlier enthusiasm for lithography.

With no knowledge of Niépce's heliography, nor of Daguerre's experimental work, the English botanist and mathematician William Henry Fox Talbot in 1833 began to explore the possibility of capturing images from nature. Like Niépce he was frustrated by the fact that he had no talent as an artist. In 1833 he tried to sketch the magnificent scenery of Lake Como in Italy using the camera lucida, which deflected a view through a mirror and lens on to the artist's drawing paper so that it could be sketched. His drawings were hopeless, however, and he got on no better with the camera obscura. In *The Pencil of Nature* published in 1844–46 he wrote:

> this led me to reflect on the inimitable beauty of the pictures of nature's painting which the glass lens of the Camera throws upon the paper in its focus – fairy pictures, creations of a moment, and destined as rapidly to fade away. It was during these thoughts that the idea occurred to me ... how charming it would be if it were possible to cause these natural images to imprint themselves durably, and remain fixed upon the paper!

Fox Talbot knew enough of science and the action of light on different materials and chemicals to work out in theory how a photograph might be created. He invented a way of printing on paper and was about to make it public when he heard of Daguerre's imaging technology, announced in 1839. In fact, Fox Talbot had discovered a completely different way of taking a photograph and one which enabled the technology to advance in entirely new directions. In 1835 he had experimented with fixing a miniature image of a lattice window and in doing so he produced the first photographic negative image in history. He made no announcement of it as it was a work in progress and he was unaware of his rivals. Talbot also discovered that long exposures, sometimes up to several hours, were not necessary as faint images could be 'developed' with chemicals applied after exposure.

As with all inventions, the art of photography advanced very rapidly once the fixing of images had proved possible. The Daguerreotype had the advantage over Fox Talbot's so-called calotype because the image was clearer. But all the prints were one-off. The advantage of Fox Talbot's method was that a negative was created, from which many prints could be made. As early as 1850, a new method of creating photographic images which combined the advantages of the Daguerreotype and the calotype was made by a retiring, little-known and unfortunate Englishman, Frederick Scott Archer.

Archer, the son of a butcher, was orphaned at a young age and raised by family and friends. Apprenticed to a silversmith and jeweller in London, he became fascinated by the sculpting of heads on coinage and studied to become a sculptor. Chiefly, he created heads and busts and took up photography so that he would have a record of his works. He found that the existing process of printing on paper did not give sufficient detail and began experiments to improve the technology. Unlike Fox Talbot, who guarded his patents fiercely, Archer simply publicised his technique, the first account of it appearing in the journal *The Chemist* in 1851:

> The imperfections in paper photography, arising from the uneven texture of the material, however much care may be taken in the manufacture of it, and which from its nature, being a fibrous substance, cannot, I believe, be overcome, has induced me to lay it aside and endeavour to find some other substance more applicable, and meeting the necessary conditions required of it, such as fineness of surface, transparency, and ease of manipulation.
>
> A layer of albumen on glass answers many of these conditions, producing a fine transparent film, but it is difficult to obtain an even coating on the glass plate; it requires careful drying, and is so extremely delicate when damp that it will not bear the slightest handling; besides these objections, the necessity of having a large stock of glass when a number of pictures are to be taken, is much against its general use. My endeavour, therefore, has been to overcome these difficulties, and I find from numerous trials that Collodion, when well prepared, is admirably adapted for photographic purposes as a substitute for paper. It presents a perfectly transparent and even surface when poured on glass, and being in some measure tough and elastic, will, when damp, bear handling in several stages of the process.

Archer's technique became known as 'wet plate' photography and for more than twenty years was used to produce the highest-quality prints even though the photographer had to carry a large amount of equipment because plates had to be prepared on the spot. The finest landscape photographs were taken with Archer's wet plates. It was discovered, too, that the quality of the images was such that it was possible to make prints smaller and smaller.

* * *

On 12 November 1870 two hot air balloons named *Niépce* and *Daguerre* rose above the Paris skyline and sailed southwards on a perilous journey across enemy lines. The French capital had been besieged by the Prussians, who had cut the telegraph lines and prevented messages getting in or out. The French Government had fled south to the town of Tours on the Loire and, as a last desperate resort to maintain contact with the captive Parisians, they had accepted the offer from the pigeon fanciers that they could use their birds to carry messages. The homing pigeons were taken out in balloons, had messages in little capsules attached to their tails and were then released on the wing and a prayer that they might find their way back to lofts on the rooftops of Paris.

There was not much in the way of conventional mail, however minutely written, that one bird could carry. One solution was to photograph handwritten messages and then reduce them. This worked, but the amount of text was still limited. And there was a much better way of cramming many messages into a tiny space. A photographer René Dragon had exhibited in Paris in 1867 his microfilm in which great landscapes, such as Niagara Falls, could be so reduced that they could be set in jewellery and then viewed through a microscopic lens.

It was Dragon and two assistants who were in the balloon *Niépce* making their escape south on 12 November. Neither balloon got far. The *Daguerre* was shot down and captured by the Prussians shortly after its launch. The *Niépce* too came down and all the equipment was lost. But Dragon and his associates managed to get away and to reach the French authorities. It took some time before the necessary microphotographic equipment could be recreated but before the end of the siege in January 1871 astonishing numbers of messages were carried into Paris.

Each microfilm measured just 3.6 x 6 centimetres and carried up to 16 pages of printed text containing around 3,000 messages. These were rolled into quills or light metal tubes and attached with waxed silk thread to the pigeons' tail feathers. The tubes, called *pellicules*, weighed less than a gram each and a pigeon could carry between 12 and 18 of them. The casualty rate for the pigeons was very high, so to ensure that messages got through they were sent in successive waves until news came that they had been delivered. In Paris the films, multiplied 160 times, were projected on to a screen where they could be easily read.

The Paris pigeon post was an epic story and, if you read the French accounts, a national triumph. However it was the invention of an English optician, John Benjamin Dancer, who was the first to use Archer's wet plate collodion photography to create miniature pictures that could be seen only through a microscope. Dancer was born in London in 1812 the son of an optician and instrument maker Josiah Dancer. When his father died he moved north, establishing businesses in Liverpool and Manchester in 1841. He was known for the quality of his microscopes, which were made for leading scientists of the day.

Dancer experimented with microphotography when the first Daguerreotypes appeared but quickly adopted Archer's collodion method, which he perfected. He produced about 500 plates in which famous works of art were reduced to such a tiny size that they could be put on the end of a pencil and in all kinds of trinkets. These became very fashionable for a while and were made popular in France by Dragon and others. It was Dragon, in fact, who patented microphotography in 1859 though he had first seen it when Dancer's microphotographs were exhibited in Paris. In a brief memoir which he dictated to a granddaughter near the end of his life, Dancer said:

> Sir D. Brewster exhibited in the year 1856 Mr Dancer's Microscopic Photographs to the members of the Scientific Academies of Paris, and one Photographer in Paris, of the name of Dragon, was not only

> impudent enough to take out a patent for Mr Dancer's invention but to commence an action against one of Mr Dancer's customers...

Sadly, Dancer's brilliant career came to an end when he began to lose his sight in 1870 after three unsuccessful operations for glaucoma. He handed his firm over to his daughters. He did have the satisfaction of knowing that his microphotography had played a significant part in the Siege of Paris and that his minuscule images of paintings were exhibited throughout Europe from royal palaces to the Vatican. And before his death the notion that microphotography could solve the problem of storing information had been discussed, although the era of microfiche and the microprocessor was a long way in the future.

* * *

At the same time as Josiah Dancer developed microphotography, the search for a way to reproduce photographs lithographically brought Senefelder's invention full circle. With official backing from the Société d'encouragement de l'industrie nationale, a number of French scientists and lithographers explored the possibility of turning photography into an industry. In particular they wanted to be able to use negatives to make lithographic prints. This involved finding a chemical formula which would fix an image that did not fade. Rose-Joseph Lemercier, who had worked as a lithographer, became interested in developing a photomechanical process in 1850 and got together with two chemists and an optician to develop what they called their 'lithographic process'. Using bitumen of Judea and the action of light, the same process that had given Niépce his first fixed image, they were able to make prints from negatives.

This relatively crude process was greatly improved by Alphonse Poitevin, who discarded bitumen of Judea and developed a more sensitive process with a chemical mix of his own. He went into business with Lemercier and soon they had photographers asking for their negatives to be printed. The new technique of photolithography was based on the discovery that exposure to light would in effect 'etch' a shape into a chemically prepared plate and that this pattern could be turned into a print. And, as an inventive young Austrian discovered many years later, into a pattern describing an electric circuit, with wires replaced by a 'printed circuit'.

* * *

When Paul Eisler graduated from the Technical University in Vienna in 1930 all he wanted was a job as an engineer working on the production of electrical goods. He was not, he recalled, especially interested in inventing or inventions. However, as a Jew in Austria he had already experienced 'baiting' by Nazi student groups and he soon discovered he could not get a job as they were all reserved for 'the virulently anti-Semitic German Nationalist Student corporations'. The British gramophone company HMV still had an office in Vienna and Eisler learned of a job going in Belgrade that would at least give him a salary. 'At the age of 23 and in a politically precarious position I looked at this offer as a lucky break.' It was the first step in one of the least known but most romantic stories in the history of invention, which in time, and by a series of extraordinary coincidences, led Eisler to combine lithography and electronics. There are echoes of the Senefelder story, for Eisler had no money and was put in a position where invention was his only hope of salvation.

The HMV job in Belgrade did not last long. Serbian railways wanted to find a way of playing gramophone music throughout their trains without excessive interference from the engine's dynamos. Eisler worked out what needed to be done, ordered the parts from England and gave a successful demonstration to railway officials. But HMV pulled out when the Serbs said they could only pay for the system in grain. Eisler returned to Vienna knowing that he would not be given an engineering job. To survive he began to work as a freelance journalist and produced a weekly radio journal. To publish details of the programmes he had to learn something about printing technology, and eventually he got a salaried job with a publishing house, which bought up his radio journal. Adolf Hitler put an end to that in 1934, for Eisler dared not associate himself with any publication that might appear to be subversive.

There was only one way out: flight to England. Eisler thought that offering inventions to British companies would give him the best chance of escaping Vienna and he came up with two, the second of which was stereoscopic television. This was very early in the development of the new technology but the Marconi Company offered him some money and paved his way to London. He needed a job but was not allowed to take paid employment so he turned to invention once again. He found a room in Hampstead, north London, and it was here that he began to work on what he would call the most significant of his inventions: the printed circuit.

In his memoir *My Life with the Printed Circuit*, published in 1989, he recalled:

> the necessary tools and implements for experimentation in this branch of engineering were sufficiently small and inexpensive to allow me to buy enough and use them in my small room. Thus it was that I turned my attention to inventions within the realm of electric circuit boards. To this I was able to bring the other practical expertise I had developed: that of printing technology acquired during the years spent in Vienna as a technical editor. . . . I learned far more after reaching Britain, by installing myself in the library of the British Museum. I became fascinated by the impressive technological achievements of the printing art. I saw this art as a whole: letterpress and gravure, lithography, offset and screen printing, engraving and photochemical printing. As I read, I imbibed all the main processes like the wisdom of the ultimate redemption.

The revelation for Eisler was that anything that could be drawn in black and white could be magnified or reduced to the size of a postage stamp and could be printed by any number of processes. The printed circuit would be a diagrammatic representation of the wires connecting component parts in a wireless or some other electrical apparatus, and it would then be printed on to a board made of suitable material. Eisler anticipated that this would be of immense value in the rapidly expanding electronics business as it would cut down on labour and allow for a reduction in the size of electrical goods.

To demonstrate his invention he made a wireless set built on a printed circuit board. Armed with this, Eisler hired an agent to find him a company that might be interested. Soon he was demonstrating his wireless to a director in the large firm of Plessey. Here he came up against the inertia of big business: they could not see the need for it. Eisler wrote that 'acceptance might discredit the firm's own Research and Development Department, or simply detract from the erstwhile dominant group in the firm, or even provide ammunition for another group in the power struggle of the firm'. He concluded that it was a myth that big companies snapped up new inventions: quite the reverse. The final put-down was that his printed circuit would do away with the girls who assembled wireless sets and that this labour was cheaper and more flexible than the technology needed to manufacture Eisler's invention.

Undaunted and ever enterprising, Eisler got a job with the rapidly expanding Odeon cinema chain, introducing his novel system of sound reproduction. At last he had a work permit and a fulfilling role as the 'ideas man' for the founder of the Odeon chain, Oscar Deutsch, who in 1936 was opening fifty 'Super-Cinemas' a year in England. Among the innovations he brought in was a way of allowing the audience to shout for an encore of a song or scene in a film as they might in theatre. If the manager thought the call sufficiently strong he would have the film wound back and the episode replayed.

Eisler did not spend long with the Odeon cinemas. Much of his time was taken up with trying to rescue relatives from Austria. His father died, a sister committed suicide, and then he was interned as an enemy alien. He was released in time to endure the Blitz in London, along with his ailing mother. Listening to the appeal made by Winston Churchill for more engineers, Eisler returned to the printed circuit with the idea that it might be useful for the war effort. He did not have any faith in the Government, which he felt sure would take no interest in his idea, and he could not raise any private capital. In the end he took a job with a long-established printing firm Henderson & Spalding, which had reputedly once published Beethoven's compositions. The main factory in south London had been bombed out but there was a basement in Shaftesbury Avenue in London's West End that served as a workroom and office. Eisler was paid £8 a week to work on a machine called a Technograph and given the freedom and funding he needed to develop his printed circuit. H. Veazey Strong, the owner of Henderson & Spalding, recognised that, whereas most printing was considered superfluous to the war effort, printing electronic circuits might not be. And he was right.

After two years' intensive development of the printed circuit Eisler had perfected a manufacturing technique and he applied for patents covering a wide range of their uses. He then, with Strong's assistance and connections, began to invite engineers and military personnel engaged in the intense technological battle with the Germans to the basement in Shaftesbury Avenue. He showed them his wireless and they discussed other applications of his brilliantly streamlined electronic circuits. But the Ministry of Supply could see no use for them. The Americans, however, very quickly recognised their potential as a key component in the 'proximity fuse', a miniature radio transmitter and receiver which could be fitted into anti-aircraft shells and other missiles.

It was very difficult to get a direct hit on a target such as a V–1 'doodlebug' flying bomb or a Japanese kamikaze dive-bomber. The use of pre-set fuses had great limitations as the shell would have to be timed to explode when it was close to the target. A proximity fuse used radio to 'sense' the presence of the target and would explode close enough to it to seriously damage or destroy it. The radio equipment housed in the shell had to be robust enough to withstand the shock of the launch. Printed circuits were ideal: compact and lightweight and strong.

The proximity fuse was regarded as one of the most successful technological innovations in weaponry made during the Second World War. Until 1948 its manufacture was regarded as classified in the United

States and so the printed circuit did not come into general use immediately after the war. But the technique of using photography to print circuits had become well established by firms such as the Diamond Ordnance Fuze Laboratories and it was just a step towards the miniaturisation of all kinds of electronic apparatus.

* * *

Shockley gave himself a whirlwind course in how to become an entrepreneur, had a look at the potential market, and went in search of a backer. Before the end of 1955 he had found Arnold Beckman, a fellow graduate of Caltech, who had created a successful business, Beckman Instruments, which had 20,000 employees around the world and headquarters in Fullerton, California. None of the Bell Labs team would be following Shockley anywhere so he began an intense recruitment drive, searching out the very best of the young talent. Anyone with a knowledge of, or background in, transistors would be a prize. Though Shockley had no knowledge of him there was one young physicist who, by chance, had had a unique insight into the first semiconductor device. If he had not stolen a piglet, Robert Noyce might just have missed out on a spectacular career.

In the late 1940s there was a fashion in America for brightly coloured Hawaiian shirts, a reaction perhaps to wartime austerity. With the shirts came a fad among students for their own version of a Hawaiian Luau party in which an animal was slaughtered and eaten, a kind of South Pacific hog roast. At Grinnell College in the farming country of Iowa, male students living in halls competed to throw the best party each year in the hope of attracting as many girls as possible. Bales of hay and other farming props were often taken illicitly for these events. In 1948 a group of young men planning a Luau decided to go one better: they would steal a pig to roast as the centrepiece of their party. The student chosen to get the pig was Robert Noyce, who knew the lie of the land as he had been brought up in Grinnell where his father was a Congregationalist minister. Grinnell College itself had been founded by Congregationalists, and their children were given the special privilege of a scholarship.

In farming country, stealing livestock was a serious offence, yet Noyce, knowing the risks, went ahead. It was in his nature: he was already in trouble because his girlfriend had told him she was pregnant. Maybe he thought things could not get much worse. But they did. Noyce got his pig and its disappearance was not noticed. Nor did the ear-piercing noise of the slaughter of the animal in a dormitory shower alert anyone to the theft. But the conscience of the boys got the better of them and Noyce owned up. It turned out he had stolen a pig belonging to the Mayor of Grinnell and the expectation was that Noyce's time at Grinnell was over and his education would be cut short.

He was saved by Grant Gale, the professor of physics at Grinnell, who knew Noyce and his family well as they attended the same church. Gale was an inspirational teacher, encouraging experimentation rather than textbook learning. And he liked to keep up with the latest advances in technology. Gale recognised in Noyce an exceptional student and fought hard to keep him. It was decided that expulsion for one semester would be sufficient punishment. In the summer and autumn of 1948 Noyce took a job at Equitable Life in New York, computing statistics in a traditional fashion – tedious, clerkish work which he hated but which he took as a penance.

It was while Noyce was in New York that Gale saw a small paragraph in the *New York Times*, published on 1 July, 1948 which was one of the few mentions of the discovery of an alternative to the vacuum tube. The head of research at Bell, Oliver Buckley, was a Grinnell graduate with two sons at the college at that time, and John Bardeen, co-inventor of the transistor, had grown up with Gale's wife and attended the same college as Gale in Wisconsin. Keen to learn more about the new device, Gale wrote to both these contacts asking if they could send a couple of transistors. Buckley replied that they did not have any to spare but he forwarded some technical papers produced by Bell. When Noyce returned to Grinnell, Gale was ready and waiting with his treasure trove of transistor papers and the two began to study them together.

It was not yet clear that the transistor would transform the whole of the electronics industry. In effect, Gale gave Noyce an early start and the ambition to find a way of making this novel technology a practical substitute for the vacuum tube. First, however, Noyce would have to get his PhD as the industry he wanted to join was being created by graduates who had doctorates in physics. He got a place at the Massachusetts Institute of Technology (MIT), taking a labouring job to pay his way through the first semester. After a faltering start in which he found he was lagging behind the other students, he excelled and won the teaching jobs and scholarships he needed to fund his studies.

Noyce, like Shockley, enjoyed amateur dramatics and he took part in a musical at Tufts University as he was finishing his doctorate. The costume designer was Betty Bottomley, a young graduate with a sharp wit Noyce appreciated. They began a relationship, with Betty typing up Noyce's thesis for him. After three months together Noyce phoned his father and asked him to marry them, though the Minister had no idea who the bride was. Noyce's biographer, Leslie Berlin, says a daughter has suggested that they married so precipitously because they thought Betty was pregnant and Noyce did not want to be responsible for another illegitimate child. Nothing was said at the time but in fact it was fifteen months after their marriage in August 1953 that their first child was born.

Noyce had to find work, and he wanted to pursue the interest in transistors he had been given by his saviour at Grinnell College, Grant Gale. He reasoned that if he was taken on by Bell Labs, who offered him a research post, he would be a very small fish in a big and exclusive pond. He rejected IBM for the same reason. Instead he chose Philco, a Philadelphia electronics firm that made radios and televisions, which he thought was a better bet as it had a relatively small team working on solid state electronics. Noyce was clear that he did not want to be lost in a large company, nor did he want to spend his time developing the theory of solid state. He wanted to make something that was useful, and the transistor would certainly be that if it could be mass-produced so that it would become a reliable and affordable replacement for the vacuum tube.

Noyce had been working at Philco for three years and had recently published a paper on the study of the surface structure of semiconductors when he got a phone call from William Shockley. He said later that it was like getting a 'call from God'. Shockley had noticed Noyce's paper and invited him to join the elite team of scientists and engineers he was putting together in Palo Alto. There was no hesitation: Noyce, Betty and their young son Billy packed their few possessions in Philadelphia and headed for California. Apart from any other considerations, Noyce said later, California was utopia for someone brought up in Iowa shovelling winter snow. One of Shockley's foibles was a belief in dubious psychological tests and all his new employees had to pass. Whatever was required of Noyce he had it, and he was soon joining a group of young physics doctorates and electronic engineers in a rough and ready factory building in which they were to subject pure silicon to a variety of tests in the hope of understanding its properties and how it could be manipulated to create a serviceable mass market transistor.

There are no really graphic descriptions of the atmosphere at Shockley Semiconductors, though an attempt to conjure up the excitement of experimentation was made in an article by the journalist and novelist Tom Wolfe in *Esquire* magazine. In 'The Tinkerings of Robert Noyce: How the Sun Rose on Silicon Valley', published in 1983, he wrote:

> Every day a dozen young Ph.D.s came to the shed at eight in the morning and began heating germanium and silicon, another common element, in kilns to temperatures ranging from 1,472 to 2,552 degrees Fahrenheit. They wore white lab coats, goggles, and work gloves. When they opened the kiln doors weird streaks of orange and white light went across their faces, and they put in the germanium or the silicon, along with specks of aluminum, phosphorus, boron, and arsenic. Contaminating the germanium or silicon with the aluminum, phosphorus, boron, and arsenic was called doping. Then they lowered a small mechanical column into the goo so that

> crystals formed on the bottom of the column, and they pulled the crystal out and tried to get a grip on it with tweezers, and put it under microscopes and cut it with diamond cutters, among other things, into minute slices, wafers, chips; there were no names in electronics for these tiny forms. The kilns cooked and bubbled away, the doors opened, the pale apricot light streaked over the goggles, the tweezers and diamond cutters flashed, the white coats flapped, the Ph.D.s squinted through their microscopes, and Shockley moved between the tables conducting the arcane symphony.

The creator of this heady atmosphere was crowned in November 1956 when Shockley learned he had been awarded the Nobel Prize for Physics jointly with Walter Brattain and John Bardeen for their pioneer work on the transistor. This was the high point for Shockley's bid to become a silicon entrepreneur: in the summer of 1957 he was devastated by what he regarded as treachery. Seven of his hand-picked physicists and engineers, infuriated by Shockley's heavy-handed style and lack of business acumen, quit when they had found a backer to underwrite a new business they would run themselves. Noyce was not one of the original rebels but he agreed to join them when he realised Shockley was finished. Though he was the last to sign up, within two years he was chosen as vice president and general manager of the new company. It was named Fairchild Semiconductor Corporation after the Fairchild Instrument and Camera Corporation, which provided the start-up funds. Shockley moved on to academia and later wrecked his public standing by promoting a eugenicist and racist view of intelligence, ending his life in obscurity.

In 1957 there was no established commercial market for transistors and none at all for solid state transistors of the kind that the Shockley rebels wanted to work on. However, they found a ready customer in IBM, which was developing for the US military a navigational computer for a long-range strategic bomber. The computer had to be both lightweight and powerful. Fairchild Semiconductor had not made any such transistors but in negotiations with IBM Noyce blithely assured them he could fulfil the order. IBM wanted one hundred of them. In February 1958, Fairchild set to work devising sophisticated pieces of equipment from scratch. In contrast to Shockley's venture the working atmosphere would be friendly and co-operative, however intense the need to fulfil the order.

Although the Fairchild team were highly qualified academics and professionals, the equipment they needed to turn silicon into a new kind of superfast transistor which could withstand high temperatures simply did not exist. There was no certainty either that they could make it work. Gordon Moore, who was responsible for the furnaces in which silicon was heated to very high temperatures, had to ship them in from Sweden. In his view Fairchild Semiconductor was able to take the huge risk it did because it was not working in a large organisation. He told the Computer History Museum in an interview in 1976:

> I think one thing that was very valuable at Fairchild in the work we were able to do was the immaturity of the organization. We did it because it was interesting and exciting. We didn't do it because we understood that there was a huge market out there that we could potentially tap. I think that the same kind of ideas starting at a GE or Westinghouse with the same amount of background and saying, 'We don't know why the heck to do it except that it looks like it's interesting' would very likely have gotten squelched as not being a very worthwhile way to go.

Although Fairchild's first customers were the US Military, and subsequently the Space Programme; they wanted at all costs to avoid government funding. Noyce had experienced it while working at Philco and he found it stultifying. In fact as T.R. Reid points out in *The Chip: How Two Americans Invented the Microchip and Launched a Revolution*, all three US services had their own fatally flawed notions of how to achieve the miniaturisation of electronics which they so desperately needed. Fairchild was able to avoid the weight of ill-informed bureaucracy, and there was no better leader than the highly individualistic, risk-taker

Robert Noyce. The team specialised in the various aspects of production and compared notes, working the alchemy Shockley had aspired to but which he wrecked by his unreasonable behaviour.

Noyce and another of the Shockley 'rebels', Jay Last, had responsibility for developing what had become known as 'photolithography', the process whereby the pattern of an electronic circuit was projected on to a semiconductor in miniature. The term had been coined by two researchers working for the National Bureau of Standards who had been given the task of making a much smaller and lighter version of the proximity fuse than those that had been made using printed circuits. Working with germanium rather than silicon, Jay Lathrop and James Nall set out to create the smallest transistor they could and chanced upon the use of 'photoresist', a photosensitive liquid, which was used to define etched rivet holes in metal aircraft wings.

Photoresist was made by the photographic firm Eastman Kodak, who obliged Lathrop and Nall with a 'quart sample'. Mimicking the creation of the wing rivet patters they got negatives of large-sized patterns and projected them through a 'microscope run in reverse'. It worked, and they showed their new creation at the Belgian World's Fair in Brussels in 1958. The following year they got a patent. According to Lathrop what they were really doing was photo-etching, but they thought *photolithographic* sounded more 'high tech' and the term stuck. With some pride he wrote: 'From these first days of semi-conductor ICs to the present, photolithography has been the primary method used by the semi-conductor industry to fabricate microelectronics.'

Exactly where Noyce and Fairchild learned photolithographic techniques is not clear. Noyce told the IEEE Global History Network:

> Some work was done in photoengraving at Philco when I was there. It had not been perfected, but the concept was beautiful. A design would be done once and one could reproduce very fine geometric patterns just by taking a picture. Certainly one of the ideas in the

> back of our minds was, 'Gee, that's a neat way to make fine structures.'

It had been experimented with at Bell Labs. In fact, Shockley had met Lathrop and Nall in Belgium in 1958 and invited them to visit him in California – an offer they declined. Both in fact joined the electronics companies at the forefront of transistor manufacture: Lathrop went to Texas Instruments and Nall to Fairchild.

Impressively, the Fairchild team fulfilled the order for IBM on time. But they encountered some serious problems with the manufacture of transistor chips. They were turning them out in batches on single silicon disks and they found that many batches had few saleable transistors. Puzzling over this, one of the team, Jean Hoerni, discovered a way of protecting the surface of the chip with an oxide layer. It then occurred to Noyce that this oxide layer could be used to cut patterns into the silicon. In fact it could be constructed with different overlapping layers, a technique they called the 'planar' process. It also occurred to Noyce that they could make the production of transistor circuits more efficient by combining several transistors on a single chip. In an interview with the WGBH radio station in the last year of his life, Noyce explained:

> We made by diffusion, photo engraving – a great number of transistors on a slice of silicon. And then those little pieces of silicon were all cut apart. And that slice of silicon was cut into little tiny pieces . . . maybe a hundredth of an inch on a side and assembled under microscopes into transistor cans. That seemed silly. Because what we would do would be to do all of the work of putting wires on those little chips of silicon – shipping them to the customer who would proceed, at that point in time, to put them all back together again. It seemed a lot easier to just simply fasten them together when they were still located on that slice of silicon.

What those in the industry called the 'monolithic idea', the notion that you might somehow embed all the component parts of an electronic chip without any wire connections, became a reality. Noyce always said he was not setting out to make an integrated circuit: it came about when they were trying to solve a problem. 'I oftimes think that the mother of invention is laziness. That you just simply don't want to go through all that work. You're trying to find an easier way of doing what it is you're trying to do,' he told WGBH.

Working quite independently of Fairchild, Jack Kilby, an engineer at Texas Instruments, arrived at a basic integrated circuit by a different route. With time on his hands when he had just joined the firm and was not entitled to a holiday, he experimented with combining in a single piece of germanium not just transistors but resistors and capacitors. He got his patent in just before Noyce filed for Fairfield and the patent lawyers fought a battle that went on for years and ended in a draw. Kilby got the Nobel Prize in 2000, and no doubt Noyce would have shared it for the invention of the integrated circuit had he lived another ten years, but the prize is never awarded posthumously.

One of the inventors of the transistor, John Bardeen, likened the creation of the integrated circuit to the invention of the wheel. Noyce told WGBH in 1990: 'I compared it at one time to the printing press. . . . you could design it once, and then reproduce it many, many times, very inexpensively. Compared to, let's say, having the monks write down the book and copy it by hand. Which was sort of the way we were building electronics at the time . . . with the integrated circuit, we got the chance of doing the whole thing identically time after time.'

Fairchild Semiconductor was hugely successful, but divisions appeared when the company had to decide if it was essentially a manufacturer of a sought-after product, the transistor, or an experimental group of scientists and engineers at the forefront of technological innovation. In 1968, to the great surprise of those in Silicon Valley, Noyce quit Fairchild. He was not yet very wealthy but he had shares to cash in and a record that was almost certain to attract venture capital. Gordon Moore, one of the original rebels, was persuaded to join him. Arthur Rock, a venture capitalist who knew them well, raised $2.5 million in two days, remarkable, as he said later, in the days before mobile phones and email. Noyce and Moore cashed in their Fairchild shares and in no time they had funding for a new company and a business plan that would make or break them. They would not be marketing integrated circuits, which were essentially 'logic chips', but would concentrate on memory. With each addition of a transistor, essentially the on–off switch used in storing information, memory increased. They could refine the technology to get more and more transistors on a chip. As early as 1964 when there were around sixty components on a single chip, Moore had remarked casually that it seemed the number would double every eighteen months. He was astonished to discover a few years later that what became known as 'Moore's Law' held true.

Noyce and Moore called their new enterprise Intel. It would be run on democratic lines and key employees would be entitled to shares in the company. Noyce made sure his old college Grinnell had a stake. Intel would concentrate on producing memory. However, the manufacturing process proved extremely difficult, with many chips discarded. The market was not too buoyant anyway. It looked at first as if Noyce had taken one risk too many. An order from Japan saved Intel, made its investors a fortune, gave rise, inadvertently, to the Altair, and brought fame and fortune to Noyce and his partners.

* * *

Noyce had become well known to electronics firms in Japan from his time at Fairchild. Those who worked with him at the time said that some Japanese electronics engineers regarded him 'as a god'. It was not entirely surprising then that he was sought out in 1969 by a young firm called Busicom, which made desktop calculators and business machines. Intel was in business to make memory chips and had a small market in creating custom-made integrated circuits. Busicom wanted them to design and manufacture some complex chips for a new and sophisticated calculator its engineers had designed, as no firm in Japan was able to do the job. It was not a major order for Intel. They were intent on perfecting the memory chip, the manufacturing process of which had not yet become sophisticated. It was a while before they discovered that the great enemy of microelectronics was contamination: the factories have to be totally free of even the minutest atmospheric particles, which might disrupt the microscopic electric circuitry.

In its early days Intel very nearly went bankrupt. It was saved initially by a deal struck with a Canadian company, which bought Intel's expertise and gave the company a cut of its profits on the sale of chips. The Busicom order was taken up by a newly arrived graduate from Stanford, Ted Hoff, one of the few Intel specialists who knew about computers. When he was at college the replacement of vacuum tubes with transistors had brought about the production of what were called minicomputers and Hoff had worked with these. Noyce knew absolutely nothing about computers: that was not his business.

Hoff looked at the Busicom design for the calculator and concluded that it would cost as much as a minicomputer and yet it would have a more limited applicability. He discussed it with the Japanese engineers who had arrived to oversee the design and manufacture of the chips they needed. They did not want to accept his judgement that it would be too expensive to produce. Hoff discussed it with Noyce, who agreed that they could not produce the chips at the price Busicom hoped for. Over time, pondering the issue, Hoff came up with an alternative design. It would be a computer on a chip, in effect. Some of the engineers in Intel thought this ridiculous. Noyce, however, was intrigued. So were the Japanese, who agreed to go with Hoff's design.

What Hoff designed was a programmable chip. Again Noyce liked the idea, as it seemed to him wasteful that all the integrated circuits they produced were custom made. Hoff's chip could be adapted to a great many uses. But it was not regarded as Intel's normal line of business. That changed when they hired a new marketing director, Ed Gelbach, who came direct from Texas Instruments. This was the firm with the rival integrated circuit designed by Jack Kilby. With encouragement from Patrick Haggarty, the visionary president of Texas Instruments, Kilby and two others had been successful in using the integrated circuit to produce the first pocket calculators, which were basically minicomputers. Gelbach brought news that Texas Instruments were now close to producing something very like Hoff's chip: in effect, a microprocessor.

Noyce began to imagine what the computer on a chip might do: traffic control, supermarket checkout, stock control. But this was all in the future. Would there be a market for these minicomputers? More importantly, could Hoff's notion be made to work? As a model for a computer Hoff had the PDP–8 he had worked with at college but transforming that blueprint into something that could be etched into a chip of silicon was no simple matter. Intel bought in expertise: the chip designer Stan Mazor joined from Fairchild and when Busicom began to put the pressure on, Frederico Faggin, another former Fairchild computer designer, was put in charge of finishing the project. Masatoshi Shima, a Busicom engineer, flew in from Japan to find out how it was going and was taken aback to discover how rudimentary it was. He stayed on for six months while all the problems were ironed out.

In the end the microprocessor, named the 4004 after the number of transistors on the chip, was incorporated into the new Japanese calculator. Busicom had exclusive use of the 4004 but Noyce negotiated a deal in which it could be sold to other companies. It went on sale for $200 towards the end of 1971 marketed as a 'computer on a chip'. It was incorporated into a 'smart' traffic light, which could sense the flow of vehicles, and began to make its way into various forms of automation as Noyce had imagined. Texas Instruments followed with its own microprocessor and the 'computer on a chip' began to develop rapidly. It was when Intel's 8080 went on sale and Ed Roberts in Albuquerque had the inspiration to make it the 'brain' of his Altair computer kit that the personal computer was launched.

When Intel learned of the Altair, Noyce and Gelbach became excited: here was a new line of business for the company. Gelbach mocked up some advertisements for his office wall: 'fully functional $300 computer'. But Gordon Moore was horrified by the idea. According to Leslie Berlin in her book *The Man behind the Microchip*, when Noyce made the casual remark 'Now that we are in the personal computer business' Gelbach recalled Moore being so incensed that 'he was going to faint or hit me'. Moore won the argument: Intel decided not to compete in this brand new market and would stick to providing components for it. This turned out to be a sound business decision, for Intel was to become one of the foremost manufacturers of microprocessors in the world, with computers proudly proclaiming that they are powered by one or other of the Intel Pentium range.

* * *

Intel became a fabulously successful company and Bob Noyce hugely wealthy. Just at this time his marriage, which had lurched from one crisis to another, fell apart. Leslie Berlin in her biography of Noyce, describes his open affair with a woman who worked for Intel and his love of intrigue: he liked to talk to his wife on the phone while his mistress was secretly next to him. In Silicon Valley in the 1970s executives talked of 'dating' women other than their wives and were not shy of showering mistresses with presents and taking them to events where their cheating would be obvious. The affair put an end to Noyce's marriage but he did not marry the woman with whom he had had such exciting times, flying her around in his private jet. Instead he married Ann Bowers who was head of personnel at Intel.

One of the ways in which Noyce retained his love of risk was in venture capital: he backed all kinds of start-up businesses and kept the details in a box. Ann Bowers also liked to dabble a little in shares. When Jobs and Wozniak were looking for financial backing for their infant Apple company a presentation to Intel was arranged. The sartorial fashion in Silicon Valley at the time was smart, or smart casual. Long-haired and hippyish, Jobs and Wozniak looked distinctly out of place and had a disconcertingly offhand manner. Noyce was not impressed: he thought the personal computer would happen but that these guys were not the ones to bring it about. His wife Ann, however, recognised that they might look like grubby students but also that they were smart:

especially Jobs. She invested, buying some shares Wozniak wanted to sell. Noyce thought she was crazy.

However, Jobs was a great admirer of Noyce and sought him out. They became good friends and, according to Ann, Jobs was a frequent visitor to their home, scheming and discussing the computer business. Despite warnings from friends Noyce remained a smoker, a two-packs-a-day man, which gave him a distinctive, gravelly voice. On 3 June 1990 he suffered a violent heart attack after he had taken his regular early morning swim at his home in Austin, Texas. He died that day in hospital at the age of sixty-two.

His memorial service in Austin was attended by a thousand mourners, and double that number gathered in San Jose, California at ceremonies organised by his brother Gaylord. Later that June on Silicon Valley's Bob Noyce Day, hundreds of red and white balloons rose into the California sky as an executive jet, which Bob had bought but never flown, shot low over the buildings in a tribute flight. George H.W. Bush phoned Bob's widow, Ann Bowers, to offer his condolences. Apple Computers paid a special tribute: 'He was one of the giants in this valley who provided the model and inspiration for everything we wanted to become. He was the ultimate inventor. The ultimate rebel. The ultimate entrepreneur.' Robert Noyce's influence was worldwide: there were services in his memory in Japan.

When asked in an interview not long before he died what he thought was the greatest achievement of his career he said, without hesitation, 'the personal computer'. Whether he knew it or not, he owed something of his success to the man who invented lithography and the weaver who first thought of the idea of programming a loom. As so often happens with inventions, Noyce had no idea not only that the microprocessor would bring about the personal computer but that it would be an essential technology in the creation of the mobile phone.

CHAPTER 5

HARD CELL

The failure of a piece of hardware had made the day long and anxious for the team working to prepare for a demonstration of the technology on which their company, Bell Telephone Laboratories, had already lavished hundreds of thousands of dollars. Not everyone at Bell believed it could work, and many of the marketing people thought that, even if it did, there would be little demand for it. But for the scientists and engineers it was a fascinating challenge. Gerry DiPiazza was in charge of a team carrying out tests in 1973 in the towns and suburbia of New Jersey, cruising the streets and country roads in a converted Barth mobile home fitted out as a laboratory and packed with sophisticated electronic equipment. Because of the technical hitch earlier in the day Gerry and his team were not in a position to check the system until 2 a.m., just a few hours before a party of executives were arriving to judge whether or not their money had been well spent.

The only way to be certain it was all working was to make a phone call from the mobile laboratory. At that time of night there was only one person outside his team he could trouble: his wife Emmy. She had been fast asleep, but Gerry needed to keep her talking while he drove to and fro across an invisible boundary between the towns of Whippany and Denville in New Jersey. After a while he asked Emmy if she had noticed any problems with reception during the call: dropouts or interference. When she said it all sounded fine Gerry mentally punched the air. This, he recalled many years later, was thrilling, because during the time he had been talking to Emmy the call had been switched between two base stations that were housed in 16-foot Prowler trailer homes. One of these stations covered Whippany and the other Denville. They were set up to work at different frequencies on the wireless spectrum so a call in one base station 'cell' would not interfere with a call in the other 'cell'. Gerry had crossed the boundary between the cells ten times as he talked to his wife and she had noticed nothing. The switch had been seamless.

What Gerry and his team had demonstrated in practical terms was what the systems engineers had called 'hand off': the ability to track a mobile phone call and to switch it from one frequency to another and on to the telephone network without the callers noticing anything. The call to Emmy was a 'eureka moment' not just for Gerry but for the whole Bell Labs team, which was working towards the creation of what would become the technology that enabled the earliest cellular mobile phone networks. Not that all the executives were impressed when they witnessed the same technological marvel that morning: one complained that the quality of sound was nowhere near landline standards. And not even the Bell Labs innovators had any notion of what this new technology would lead to. In their minds it was a way of allowing many thousands more in American cities to use a car phone without interfering with each other's calls. Sometimes they joked among themselves that a time would come when someone had their phone number embedded in their head at birth and if they did not answer you knew they must be dead. But nobody in the 1970s foresaw the spectacular spread of the miniature mobile phone across the world that began in the 1990s.

At more or less the same time as Gerry DiPiazza experienced the excitement of the working 'hand off', Bell's rival, Motorola, which had a large share of the radiophone market in the United States, unveiled the world's first handheld radio telephone, which they called the

DynaTAC. Whereas the Bell breakthrough had been strictly in-house, Motorola wanted the press and the politicians and authorities in Washington to know about it. A series of demonstrations was staged in New York and Washington and Motorola confidently predicted that a full DynaTAC system would be in operation covering the whole of New York by 1976. Vice President John Mitchell was quoted in a Motorola press release as saying:

> What this means is that in a city where the DynaTAC system is installed, it will be possible to make telephone calls while riding in a taxi, walking down the city streets, sitting in a restaurant or anywhere else a radio signal can reach. . . . We expect there will be heavy usage by a widely diverse group of people – businessmen, journalists, doctors, housewives, virtually anyone who needs or wants telephone communications in areas where conventional telephones are unavailable.

In the same press release, Vice President Martin Cooper, in charge of the team that produced the handheld mobile phone, explained: 'As the portable phone user talks his voice is transmitted over the air much the same way as a two-way radio station transmits. This message is picked up by a receiver, relayed to the DynaTAC central computer and fed into the regular phone network. . . . As a portable phone user moves about the city in the midst of a call the computer will switch the conversation to different transmitters and receivers as required to assure continued clear message quality. This happens so quickly neither party in the conversation is aware of it.'

Mitchell and Cooper had a vision of the mass ownership of what *Popular Science* magazine called in 1973 'take along phones' when many in the industry were sceptical about the demand for such a novelty or its technical feasibility. A good deal of research was still needed to make it a commercial proposition and the whole enterprise was dogged in the United States by political prevarication. This allowed Japan and an alliance of Finland and Scandinavian countries working as Nordic Mobile

Telephone (NMC), to beat the American pioneers to the introduction of the first commercial cellular networks. In fact, in a rather bizarre turn of fate, the world's first fully functioning cellular mobile system for car phones was created by the Swedish Ericsson and Dutch Philips in Saudi Arabia in 1981 as a luxury add-on to a recently installed landline system.

The story of the creation of the cellular mobile phone has barely been touched upon by historians. Whereas the inventors of the two technologies that it brought together – the telephone of Alexander Graham Bell and the wireless of Guglielmo Marconi – are well known, there is nobody to whom it is possible to attribute the creation of a fully functioning mobile system. However, within the communications industry there are those who have been awarded prizes by their peers for the exceptional contribution they made in the pioneer years of the 1960s and early 1970s. Martin Cooper of Motorola is one and he has shared the honours over the years with some brilliant engineers who worked in Bell Laboratories, the extraordinary research establishment of AT&T that was founded on the fantastically lucrative patents of Alexander Graham Bell.

Whereas in the development of many modern inventions the amateur has played a significant part in the scientific and technological discovery of something useful, the creation of a cellular telephone network was far too complex and costly an endeavour for the tinkering individual. Nevertheless, at Bell Laboratories individual engineers and scientists would routinely spend some time mulling over ideas that were outside the mainstream of their research. The problem of how to create a network that could handle many more wireless phone calls than existing technology was a passing interest at Bell from the late 1940s. A proposal for how this might be achieved was written as early as 1947, but it was futuristic for two reasons. First, wireless frequencies were strictly controlled by the Federal Communications Commission (FCC) and allocated between competing users, and for the Bell scheme to work AT&T would need to be awarded a new slice of the spectrum. And secondly, even if the additional frequencies had been allocated, the complex system of exploiting them that Bell engineers had dreamed up would require technologies yet to be invented. As with all innovations, the creation of what is now the familiar mobile phone network required a piece of equipment which had nothing to do with telephones. Just as the powered aeroplane needed the petrol engine to get off the ground and television the amplifying valve to boost the signal from the photoelectric cell, so the mobile phone had to await the microprocessor or minicomputer.

The first impetus to explore the cellular idea came from the FCC, which decided that a slice of the wireless spectrum that had been allocated to small TV stations was being wasted. In 1968 the Commission called for ideas about how this scace resource might be better used. The idea for a cellular phone system was resurrected at Bell Labs. A wireless engineer, Phil Porter, and a mechanical engineer, Richard Frenkiel, had worked on a design for a cellular system in 1966, before dropping that work to fit a phone system on a new high-speed train between New York and Washington. They, along with a young supervisor, Joel Engel, got back to work on the cellular phone concept in 1968, which proved to be infinitely more challenging (both technically and politically) than anyone imagined.

By 1971, when a detailed proposal was delivered to the FCC, Bell Labs had more than a hundred engineers working on the system, and their numbers would continue to grow throughout the decade. A functioning system with 1,000 customers was demonstrated by AT&T in Chicago, in 1978, but the political issues of a potential new AT&T monopoly, and opposition to this (particularly by Motorola), delayed commercial service until 1983, by which time there were operating systems in several other countries.

There were mobile phones in the 1960s, nearly all of them a luxury accessory in a car with a conventional handset next to the driver and the heavier wireless equipment in the boot. In his memoir *Cellular Dreams and Cordless Nightmares* Richard Frenkiel provides an amusing account of the old system he and his colleagues were to consign to history:

> Those early mobile telephone systems were like broadcast radio or television stations. They used a powerful transmitter on top of a tall

building in the center of a city, from where it could cover more than a thousand square miles. Only the most modern of these systems allowed subscribers to dial and receive calls automatically – many still used mobile operators to place calls. Subscribers would push buttons looking for an idle channel, or listen-in on calls – waiting to grab the channel the second a conversation ended. One didn't discuss one's business secrets, love life or criminal intentions on those systems, and conversations were carried out in a sort of code ('I'll send the proposal to that guy you spoke to . . .). Demand for service was high, but there were only a few channels in any city, and each channel could be used for only one conversation at a time in the entire service area. Imagine a system covering New York City and its suburbs, in which there are only a few hundred customers and in which only a dozen calls can be made at any one time!

Because demand was high and so few channels were available, about sixty customers shared each channel. . . . As a result, all channels were busy most of the time and it was difficult to make a call. . .

It was said that Lyndon Johnson, as a senator, repeatedly called a political rival from his car to demonstrate that *he* had a car telephone. When his rival finally pulled enough strings to get his own car phone, he called Johnson's car and said, 'Lyndon, I'm calling from my car.'

'Just a minute,' Johnson was said to have replied. 'There's a call coming in on my other line.'

That last is perhaps an apocryphal tale. However, there is an authenticated story which nicely defines the official view of mobile phones in Britain in the immediate post-war years, related in Jon Agar's *Constant Touch: A Gobal History of the Mobile Phone.* In 1954 an aristocrat, the Marquess of Donegall, learned that the Duke of Edinburgh had a mobile phone fitted in his car. It was not connected to the landline network but he could call Buckingham Palace via a relay station in Hampstead, north London. Not to be outdone, the Marquess asked if he could have one. All telecommunications were administered by the Post Office, which wrote to him in the following terms:

> There is . . . no arrangement for private persons to fit radio in their own vessels or vehicles for communications with the public telephone network. The prospects of starting such a service in the United Kingdom in the present state of technical knowledge in the radio field are nil. There is no room in the radio frequency spectrum.

It seemed, then, that a technology that began with one simple observation which thrilled the scientific community in 1820 and made possible an entirely new electronic world had finally been played out. In fact, it turned out, these were just the pioneer days.

* * *

In 1820 a report by a distinguished Danish professor excited the scientific world throughout Europe almost as much as if he had solved the old riddle of how to turn base metal into gold. Unsure which language he should use to publish his revelation, he chose Latin. The short paper was quickly translated, and it appeared in vol. 16 of the *Annals of Philosophy* of 1820 as Article IV with the title 'Experiments on the Effect of a Current of Electricity on the Magnetic Needle'. The author was already well known, his first name Hans anglicised in the *Annals*, so that he became John Christian Oersted, Knight of the Order of Danneborg, Professor of Natural Philosophy and Secretary of the Royal Society of Copenhagen. In just four pages he described a series of experiments he had made, with a party of leading scientists as observers, on the effects of an electric current on the movement of the magnetic needle of a compass.

A bank of twenty copper troughs containing zinc plates in an acidic solution provided what he called his 'great galvanic force', which flowed through a wire that he could manipulate around the compass. He described how he ran an electric current at various angles to the compass needle, at different distances from it, and repeated this many times, sometimes with the compass completely enclosed in a variety of containers. What he showed conclusively was that around his charged wire there was some kind of magnetic force, which, according to the direction of the current, could attract or repel the needle. The effect weakened when the wire was moved further from the compass, but enclosing the compass did not alter the response of the magnetic needle.

Oersted had first suspected that electricity and magnetism might be intimately related in some way when he noticed during a lecture a slight movement of a compass needle which lay close to a charged wire. It was an observation to which he was receptive as a devotee of the philosophy of Immanuel Kant, who postulated that the world was a single organism. The scientific wisdom of the day, however, was that electricity and magnetism were distinct and unrelated forces. The American Benjamin Franklin had noticed the effect of thunderbolts on magnetic needles but had not guessed that there was an inherent relationship between the lightning and the magnet.

No sooner had Oersted's paper been translated into a variety of languages than the laboratories of leading chemists and scientists in Europe began to repeat his experiments and to debate the nature of the invisible force that was emitted by an electrified wire. There was no theory to explain what was going on and there were inevitable disagreements. In 1821 Oersted embarked on a triumphant tour in which he was lauded for his great discovery while at the same time arguing fiercely with scientists in France, Germany and Britain about how it might be incorporated into scientific theory.

In London, Oersted's arrival in May 1823 caused great excitement and it was here, in the Royal Institution, that his observations, and their examination in Europe, led to an entirely new technology without which there would be no telephones or any other of the familiar electronic gadgets of the modern world. When Oersted's paper first appeared in the *Annals*, Sir Humphry Davy had immediately repeated the experiments and had begun to take them further. Davy, newly elected as President of the Royal Society, was the leading light in the Royal Institution, which had been established in the late eighteenth century as a research centre funded by wealthy landowners and industrialists who hoped it would provide them with innovations of value to industry and agriculture. Davy published papers on his electromagnetic experiments and, on Oersted's initiative, had become a foreign member of the Royal Danish Society. Oersted, in turn, had been made fellow of the Royal Society in London.

Shortly after he had crossed the Channel and enjoyed a coach ride on turnpike roads 'as smooth as a dance floor' (so he wrote), he was entertained by the Royal Society Club at the Crown and Anchor in the Strand and later that day at the Society's hall in Somerset House. All the great names were keen to show him the delights of London, its central streets lit brilliantly with gas lamps. The astronomer John Herschel, who compared Oersted's determined exploration of electromagnetism with Columbus's discovery of the New World, took him in a rowing boat on the Thames. There had been a spate of bridge-building on the river and Herschel explained with some pride how the wide arches that allowed shipping through were supported on cast-iron spans.

There was a great deal to admire in London but nothing was more pleasing to Oersted than an intriguing performance of a magical instrument called the Enchanted Lyre. The inventor was a young man called Charles Wheatstone. He would welcome an audience into a room in which chairs were arranged with the focus of attention on a lyre, a stringed instrument rather like a harp, suspended from the ceiling by a metal wire. Wheatstone would begin the performance by appearing to wind up the instrument, which would miraculously begin to play a piece of classical music performed by several instruments. If a curious member of the audience went close to the lyre the illusion was even more telling: music was coming from it.

In fact the Enchanted Lyre was merely a sounding board: the music emanating from it was generated by vibrations in the brass wire from which it was suspended created by three musicians playing piano, dulcimer and harp in the room above. However, Wheatstone was not a mere entertainer. In fact he was an intensely shy man incapable of saying more than a few words in public. He came from a family of musical instrument makers and the Enchanted Lyre was an illustration of a subject that fascinated him: the nature of acoustics. It is not known exactly when or how Wheatstone and Oersted first fell into conversation in London but they discovered very quickly that they had many interests in common and had conducted similar experiments analysing the nature of sounds. At the time Wheatstone was not part of the scientific community and it was Oersted who persuaded him to publish some of his research, which Oersted then took with him to Paris. They stayed in touch and Oersted continued to promote Wheatstone, who was twenty-five years younger than him and just embarking on what would be a remarkably inventive career. It is likely that it was Oersted who put Wheatstone in touch with Michael Faraday who was working as laboratory assistant to Humphry Davy at the Royal Institution and these two remarkable, self-taught experimental scientists between them were to transform Oersted's revelation of the magnetic power of an electric current into the basis of all modern communications. Faraday was more than ten years older than Wheatstone and the story of his elevation from a mundane trade to the highest peak of scientific fame was the more remarkable, its twists and turns worthy of a Dickensian adventure: *Great Expectations* would not be a bad title.

* * *

In 1805 there appeared in London bookshops an anonymous work with the curious title *Conversations on Chemistry*, with the subtitle *Intended More Especially for the Female Sex.* The book took the form of a stilted dialogue between a Mrs B and two young women, Caroline and Emily, who begin by expressing surprise that anything to do with chemicals might be interesting. They are taken step by step through a few elementary discoveries and are shown some experiments. The book proved to be very popular and went into many editions, a large number of which were plagiarised in America. After a year or two, the anonymous author revealed that she was Jane Marcet, the daughter of a wealthy Swiss merchant and banker who had settled in London, and the wife of a Swiss doctor whom she had married in 1799.

Her maiden name was Haldimand and she had had unusually stimulating upbringing as the only surviving daughter of a family of twelve children. As a girl she read widely and was encouraged to take an interest in what her brothers were reading and discussing. Her mother died when Jane was just fifteen and she took over the running of the household. At the age of thirty she married Alexander Marcet, and they had four children. Marcet knew many of the leading thinkers of the day; Jane liked to listen to their conversation and began to form the idea of writing everything down. When Humphry Davy began his lectures at the Royal Institution she watched his experiments with a keen interest and determined to write a book that she hoped would spark an interest among women.

As it happened, *Conversations on Chemistry* was the book that launched the career of Michael Faraday. The son of a Yorkshire blacksmith and a farmer's daughter who had moved from the north of England to London, Faraday was working as an apprentice bookbinder when Marcet's chemical treatise came into his hands. Though he was an avid reader and was encouraged in his self-education by his master, the bookseller George Riebau, he declared later in life that no book had had such a great influence on him as *Conversations on Chemistry*. In particular it gave him an appetite for experimentation, which he began to practise in a small way in a room at the bookseller's.

Born in 1791, Faraday was the third of four children of a family that belonged to a tiny Nonconformist sect called the Sandemanians, founded in Scotland in 1730, which believed that the Church should not be involved with the State and that it was wrong to seek wealth. They were a close-knit group and Faraday remained a Sandemanian all his life. (The sect was named after an early leader, Robert Sandeman.) In London Faraday played in the streets like other children and attended a local school where he learned to read and write. At thirteen he got a job as an errand boy taking papers to and from the bookseller George Riebau whose shop was in Blandford Street close to where the Faraday family lived. Recognising the boy's ability, Riebau offered Faraday a seven-year apprenticeship as a bookbinder. This was the first of a number of lucky breaks in Faraday's life – not because he enjoyed bookbinding or the book trade, which he found tedious, but because he began to read the books in Riebau's shop.

By the time he had served his apprenticeship Faraday had compiled and bound a small collection of notebooks. Riebau allowed him to go to lectures at the City Philosophical Society, which met at the home of a silversmith, John Tatum, in Dorset Street not far from the bookshop; one of Faraday's older brothers paid the fee of one shilling. There he made some lifelong friends, including Richard Phillips, a chemist who was to play an important part in his scientific career. In 1812 a customer of Riebau who had learned of Faraday's interest in chemistry gave him tickets to four lectures by Davy at the Royal Institution: the last Davy was to give there, as it turned out, for his marriage to a wealthy widow freed him from having to earn his living.

It was also the year when Davy became Sir Humphry, knighted for his services to science. He was now grand and wealthy and still keen to continue his researches, which he did as an honorary professor at the Royal Institution with the title of Director of the Laboratory. Faraday, by contrast, had begun work full time as a bookbinder for a new employer whom he loathed. In desperation, he wrote a letter to Joseph Banks, the very grand secretary of the Royal Society, asking how he might become a scientist, and left it with the porter at Somerset House. He was rebuffed: he called many times for an answer, only to receive a note to say his letter required no answer.

Faraday had taken to making notes of lectures at the City Philosophical Society and he had done the same when he was in the audience for Davy's last lectures. At the suggestion of the man who had given him the tickets to the lectures, Faraday edited the notes and sketches he had made, bound them and delivered them to Davy with a letter to the effect that he hated 'trade' and longed to be a scientist. Clearly intrigued by Faraday, Davy met him and gave him some kindly advice along the lines of 'do not give up the day job' and was amused by the young man's naïve belief that science was an occupation superior to that of bookselling.

It was then that a rapid sequence of quite unforeseeable events transformed Faraday's life. Davy, on his honeymoon in Scotland enjoying some fishing, got news that a French chemist had used chlorine, a element he regarded as his own, to make an explosive and had injured himself. Anxious to confirm this discovery, Davy cut short his trip to Scotland and went to a laboratory in Tunbridge Wells where, with another chemist, he repeated the very dangerous experiment. A glass container blew up and a splinter damaged one of Davy's eyes. In need of some assistance, he called in Faraday on an occasional basis. Evidently pleased with him, Davy recommended him for the job of laboratory assistant at the Royal Institution when a vacancy arose unexpectedly: a previous assistant having been fired for violent behaviour.

Now not only could Faraday leave bookbinding for good, but he was soon embarking on trips to meet all the great chemists and philosophers of Europe, accompanying Davy and Davy's aloof wife, who wanted him to eat not at her table but with the servants, a snub her hosts would not allow. With Davy, Faraday was a servant when they went hunting – he was asked to load Sir Humphry's gun – and then a laboratory assistant when experiments were performed. But this was no matter to Faraday, for this wonderful trip was his university and one in which he met in Milan the ageing creator of the first practical battery in 1800, which made Oersted's experiments possible by providing a means to create a continuous flow of electrical power. Alessandro Volta was, Faraday wrote, 'a hale elderly man . . . and very free in conversation'.

By 1821 Faraday was established at the Royal Institution, where he had rooms and was made acting superintendent of the building. He married Sarah Barnard, the daughter of a silversmith who was, like his own parents, a Sandemanian. Faraday reasserted his faith, promising to live like a good Christian. Like so many scientists he saw no conflict between his religious beliefs and his research. He continued to work with Davy but also began to conduct his own experiments. Oersted's discovery stimulated immediate research and Faraday began to unravel some of the mysteries of electromagnetism. But he was obliged to take on other work which he felt was of much less significance. It was in 1831, in fact, ten years after Oersted's revelation, that Faraday discovered two vital attributes of electricity. One was induction: that a current running in a wire could induce a current in an adjoining wire; and – of even greater significance – that it was possible to generate an electric current with a magnet. His reasoning was straightforward: if, as had been shown repeatedly, an electric current could turn a piece of iron into a magnet then should it not be possible to reverse the process? It took many attempts to discover how this could be achieved, the key being movement: the magnet had to move inside a wire coil to set the current going. In this way he made the first dynamo or electricity generator, thus providing the world with an entirely new source of power.

Faraday had succeeded Davy as director of the Laboratory at the Royal Institution in 1825 and it was in that year that he became friendly with Charles Wheatstone whose scientific career had been promoted by Oersted. Like Wheatstone, Faraday was interested in the nature of sound, and in the Friday lectures he inaugurated at the Institution he delivered some of Wheatstone's papers for him. So timid was Wheatstone in public that there is a legendary story that he fled the lecture theatre, abandoning his audience and that subsequently all speakers were locked in before they went on stage in case they 'did a Wheatstone'.

Wheatstone had a very wide range of interests and was more of a practical inventor than Faraday, though his researches in all fields were just as meticulous. He invented the stereoscope, which produced 3D pictures, a variety of electrical generators, and a clock that could tell the time by the sun even on a cloudy day. But what he is best known for is the part he played in the creation of the first commercially operational electric telegraph in history. There had been a number of attempts to transmit messages with electricity before the discovery of electromagnetism and some of them worked, though they were hardly practical. Wheatstone was, by 1834, recognised as a gifted scientist and had been made Professor of Experimental Philosophy at King's College, London, which had been established in 1829 and was to become one of the two founding colleges of London University. From 1830 he had taken an interest in electricity and had been devising methods of measuring the speed of a current. He especially wanted to measure this in a long wire and got permission to suspend four miles of cable in the vaults below the King's College building.

Wheatstone had contemplated the creation of an electronic messaging system but had not sought to develop it until he was approached in 1837 by a former lieutenant in the East India Company army, William Fothergill Cooke, who, despite his lack of scientific training – he had studied classics at Durham and Edinburgh – was convinced that an electric telegraph would be a commercial success. Cooke had planned to take up anatomical modelling and was attending lectures in Heidelberg when he saw a demonstration of a small-scale telegraph system. When he returned to England he tried to construct his own version but had a problem getting a signal to travel very far. He consulted Faraday, who said he thought the telegraph was a good idea but did not take it any further. He then approached Peter Roget, secretary of the Royal Society, who suggested he consult Wheatstone, who was experimenting with his four-mile wire at King's College.

* * *

Cooke took the view that the idea for the electric telegraph was his alone and that he was consulting Wheatstone for a bit of scientific know-how. Wheatstone thought otherwise. However, they came to an agreement that, if their venture proved profitable, Cooke should get 10 per cent for acting as commercial manager and the remainder should be divided equally between them. Correctly, Cooke believed the new steam railways would be the first customers and he had already consulted the company running the Liverpool and Manchester Railway, which had opened in 1830. Wheatstone's understanding of electricity, on the other hand, enabled him to solve the problem of sending a signal a long distance. The key had been provided by the German physicist Georg Ohm, who discovered a formula that made it possible to calculate how far a current could travel in a circuit. This was published in German in 1827 and Wheatstone's grasp of the language gave him a head start on others working on the problem. In addition, he devised a telegraph mechanism in which an electric impulse jerked a needle, which then pointed to a letter of the alphabet. Although he was a brilliant cryptographer himself, Wheatstone assumed customers for his telegraph would want to be able to send and receive messages without having to learn a code.

Cooke and Wheatstone got their first patent in 1837 and installed the first commercial system on a section of the Great Western Railway between Paddington and West Drayton in 1838. This pioneer telegraph proved to be over-elaborate and expensive as it used a multiplicity of wires. But it worked and it was the first of many devised by the partners, Cooke acting as business manager and Wheatstone as technical developer before they fell out over the issue of who had invented what.

Although the first commercial market for the electric telegraph in Britain was provided by the rapidly expanding railway network, Wheatstone also aimed to attract individuals who might want to communicate at a distance. In this he anticipated the private use of the telephone. He devised a telegraph that printed messages, and to build and promote this invention in 1861 he established the Universal Private Telegraph Company. For £4 a mile subscribers could commission their own private telegraph line. There was no need for an operator as there was with all other telegraph systems, as the mechanism Wheatstone devised required no special skill in reading codes or ciphers. For commercial purposes, the sending and receiving was far too slow, but the system proved to be both popular with wealthy individuals and profitable for Wheatstone. Overheads were minimal, as a line needed little maintenance and no operators other than the owners. In the eight years it was in commercial operation the Company had laid out 2,500 miles of wire and installed 1,700 sending and receiving instruments. It would certainly have expanded more quickly had not the British Government brought all telegraphs into national ownership and handed them over to the General Post Office in 1869. Wheatstone received £9,200 in compensation, perhaps £500,000 at today's prices. The lines began to disappear when the telephone arrived a few years later, though the last private telegraph line survived until 1950.

Cooke and Wheatstone were among the pioneers of the practical, commercial electric telegraph but they did not devise the system which was to sweep aside all others from the mid-nineteenth century onwards. This was conceived on the other side of the Atlantic or, if the self-styled inventor is to be believed, in mid-Atlantic aboard the sailing ship *Sully* on its month-long voyage from the French port of Le Havre to New York.

* * *

Of all the rank amateurs who by lucky chance and determination have turned advances in scientific understanding into momentous inventions, surely none can compare with the American painter Samuel Finley Breese Morse. He had no understanding of electricity or electromagnetism and none of the mechanical skills which talented amateurs often possess. At the time he first conceived of an electric telegraph he was totally and blissfully ignorant of the experimentation of hundreds of others over more than a century that had made this a possibility. And a close study of how the Morse telegraph came about raises many doubts about his personal contribution to the final and spectacularly successful version of the innovation. Yet he is one of only a handful of inventors who achieved international fame and whose name still resonates.

Born in 1791 in Charlestown, Massachusetts, Morse was raised as an evangelical Calvinist. He was sent away to school at the age of eight and entered Yale University when he was just fourteen years old, graduating five years later. Though he was given some instruction in science, his ambition was to be an artist and in 1811 he sailed to London. Here his terracotta study of Hercules was awarded a gold medal by the Society of Arts and he exhibited some ambitious paintings, but when he returned to Boston in 1815 he found that his art was not in demand. To earn a living he took to portrait painting, wintering in South Carolina, and by

1823 was able to set himself up in a studio in New York. He was becoming established as a figure in the world of American fine arts when tragedy struck. In 1825 his wife died, leaving him with four young children. Within two years both his parents had died and his children were being cared for by one of his brothers. Abandoning them he left for Europe in 1829, where he lived in Paris and Italy. On 1 October 1832 he set sail from Le Havre on the *Sully* taking with him his painting, *Gallery of the Louvre*, hoping it would make a good impression in New York.

There is no evidence, or even a suggestion, that Morse had given anything electric even the slightest consideration before the sails of the *Sully* filled and he and his twenty-six fellow passengers were at sea. A small group got into conversation and the possibility of using electricity to send messages was discussed. This was almost certainly a topic first raised by a young doctor and scientist Charles Thomas Jackson who had been in Europe for three years studying mineralogy, geology and medicine. He had seen a demonstration in Paris of an electric current sent around a lecture theatre and he knew something about electromagnetism. Morse became fascinated and excited by the idea of creating an electric telegraph system. Facsimiles of notebooks he kept on the trip show crude drawings of bits of electrical apparatus which can only have been suggested to him by Jackson.

In New York, Morse once again found there was little interest in his paintings. He took unpaid work as a professor of sculpture and painting and in 1835 became Professor of Literature of the Arts and Design. He earned very little and was broke. His one hope of financial salvation was the idea he had for the electric telegraph. He constructed a crude and impractical recording device in which a pencil was induced to make marks on a piece of stretched canvas. His feeble electric current would send a signal only a few yards. He consulted the Professor of Chemistry, Leonard Gale, who knew something about electromagnetism and who was able to improve the signal. Most usefully, he suggested that Morse contact Joseph Henry, who was an American equivalent of Michael Faraday. Like Faraday, Henry had been raised as a devout Christian, in his case a Presbyterian. His family on both sides was of Scottish descent and his parents were settled in Albany, the capital of New York State, when Henry was born in December 1797. William, his father, was an impoverished day labourer who found it difficult to support the family and Henry was sent at the age of seven to live with his maternal grandmother. Two years later his father died, but he stayed with his grandmother until he was fifteen.

Henry recalled his first introduction to reading. A pet rabbit had escaped and sought refuge in the village meeting house. Henry crawled under the floor in pursuit and discovered a world of books, which included *The Fool of Quality* by Henry Brooke, a cautionary tale in which a pampered son fails and a rejected son triumphs. This gave him a taste for reading fiction and when he returned to Albany to live with his widowed mother, he fancied himself as an actor on the stage. He produced a number of plays, devising stage effects, and generally enjoyed the lively thespian culture in Albany. He was apprenticed to a watchmaker and silversmith and over two years acquired some mechanical skills. But it was a book left by a lodger in his mother's house that inspired his interest in science: it was an epiphany to match Faraday's *Conversations on Chemistry*. Henry's book was *Popular Lectures on Experimental Philosophy, Astronomy and Chemistry, intended chiefly for the use of students and young persons* by G. Gregory, who was a rector in West Ham, then on the fringes of London. Henry was sixteen when he read it and later inscribed on a blank page: 'This book, although by no means a profound work, has, under Providence, exerted a remarkable influence on my life. . .'

Henry enrolled at Albany Academy night school to study geometry and mechanics; he became tutor to a landowner, General Stephen Van Rensselaer, who was president of the Academy; and he assisted Dr Beck, principal, with chemical experiments. To regain his heath after years of intensive study he took a job as an engineer on a road survey between West Point and Lake Erie. He might have continued as an engineer had he not been given the chair of mathematics at Albany Academy in 1826. It was some years before he began to take an interest in the electrical discoveries which had so excited Europe and was struck by the fact that no American had pursued the subject since Benjamin Franklin. Henry proceeded to make up the ground very quickly so that by the time Morse was musing about the electric telegraph he had begun to conduct the same experiments as Faraday. In particular he had studied the electromagnets devised by William Sturgeon, another self-taught English electrical engineer – his father was a Lancashire shoemaker and notorious night poacher – and improved on them. Their power was gauged by the weight they could lift when a current flowed through a binding of wires around a soft iron bar bent into the shape of a horseshoe.

Henry had demonstrated an electric telegraph system in a lecture theatre and he was able to provide Morse with all he needed to know about sending a signal over a long distance. Whereas Morse was avid in his pursuit of exclusive rights to his alleged inventions Henry refused to patent anything as he thought it beneath the dignity of a natural philosopher. He would later get dragged in as a reluctant witness to give evidence of his own telegraph experiments for rivals to Morse who wanted to dispute his claims to precedence.

In 1837 Morse was startled to learn that an electric telegraph was already in operation in Britain. His first thought was that someone had stolen his idea. So he wrote to the captain of the *Sully* and four of his fellow passengers who had been on the voyage five years earlier when he had first conceived of the invention. Gratifyingly, they assured him that they recalled his discussing the idea of a telegraph. However, he did not contact the man who had provided him with the first rudimentary idea of how it might work: Charles Jackson. As soon as Morse began to publicise his concept of the electric telegraph Jackson fired a broadside claiming that he had given Morse the idea.

As it was, the equipment Morse had developed by that stage was a long way short of a practical telegraph system. He demonstrated it in a lecture theatre at New York University in September 1837, sending a signal in a wire about a third of a mile long to activate his cumbersome recorder. At this stage Morse Code was an unwieldy concept: a huge dictionary was being compiled in which every word had a number.

What impressed those who witnessed this demonstration was the relative simplicity of Morse's system compared with those in use in Britain. With a single wire and the message recorded rather than read (as in the Cooke/Wheatstone version at that time) Morse had arrived at a winning formula. But he lacked the mechanical skills to turn it into a practical and commercial system. The problem was solved for him by a young man called Alfred Vail, a former pupil of Morse who had recently graduated from the university. They attended the same Presbyterian church and knew each other reasonably well. Vail became intrigued by the telegraph and suggested he could help develop it. This was manna from heaven for Morse. Vail's family owned a successful steelworks at Speedwell in New Jersey. Alfred had worked there as a mechanic. A deal was struck whereby Alfred's family firm would fund the project, he would be the mechanic and they would share in the profits if it were successful. Everything, however, would be attributed to Morse whether or not he had played any part in devising it.

It took seven years for Vail and his assistants at the Speedwell Iron Works to produce a much-improved Morse telegraph system. The numbered dictionary was jettisoned and the familiar dots and dashes introduced, some say by Vail, though he chose not to claim this himself. Finally, on the historic day of 24 May 1844, Morse tapped out the portentous message 'What hath God wrought!' from the Supreme Court Chamber in Washington to a line stretching to Baltimore where, from just outside the town, Vail received the message and sent it back.

Morse by this time was well aware of the prior existence of a form of the telegraph in Britain and Europe. He had sailed to London in 1837 in an attempt to patent his system there. His application was turned down on the grounds that it had already been published in a technical magazine. He met Wheatstone, whom he judged to be 'a genius' and very likeable. And on 28 June 1838 he witnessed Queen Victoria's Coronation, remarking: 'Is it possible . . . that the English are children, to be duped by these gew-gaws?'

Though he failed to get anywhere in Britain at that time, once his system began to spread rapidly across America it took hold in Europe too. The debt to earlier experimentation was acknowledged, with one extraordinary omission: in a book Alfred Vail published in 1845 on the history of the electric telegraph. In the years he was making the Morse system work he clearly did his historical research. But he failed to make more than a mention of Joseph Henry. It was a mean-spirited snub for which Morse was most likely responsible: he had a quarter-share in the proceeds of the book and probably read it through, though much of it would have been technically beyond him.

Whatever the truth about Morse's contribution to the telegraph system to which his name is attached, its huge success was the first great transformation of long-distance communication and it provided one very significant historical development leading to the telephone. As the demand for telegrams increased at a tremendous rate the system was threatened with stagnation. A way had to be found of sending more than one message at a time down a line. It was not obvious how this might be achieved. However, a solution was found from an apparently unrelated line of research, which had to do with a study of acoustics and the nature of sound and hearing.

* * *

In 1845, after the work of half a lifetime, a German émigré, Joseph Faber set up in Philadelphia his extraordinary talking machine, hoping that he would attract more attention than he had in Europe or New York. It was one of those nineteenth-century creations that was intended both as a form of entertainment and as a demonstration of scientific principles. Faber's invention resembled a ventriloquist's dummy with a modelled head and torso decked out in a form of Turkish garb with the intention, perhaps, of lending it a magical Eastern aura. This talking head could converse in any language spoken to it: it could sing, and laugh and behave in such an astonishing way that there was a widespread belief that it was a hoax. There had been numerous bogus automatons on show before, all of which were given voice by a concealed dwarf.

For reasons nobody has ever been able to explain, Faber's life's work – it took him seven years to get his Turk to enunciate an 'e' sound – did not appeal to the general public. It attracted little more interest from the scientific community. However, it did catch the attention of Robert Patterson, director of the US Mint in Philadelphia and he thought it worth investigating. Before he could study it more closely, Faber, frustrated by the poor response to his automaton's performance, smashed the whole thing up and burned it. Patterson offered to raise money to pay for it to be rebuilt. Faber rejected the offer and set about reconstructing it with his own funds. When it was completed, Patterson arranged for Joseph Henry to take a look at it and to confirm that there was no hoax involved. Henry discovered that the talking Turk was a very sophisticated version of a recognisable kind of talking machine which had been exhibited in Europe since the late eighteenth century. It was not the first he had seen. Charles Wheatstone had created one which had impressed him when he visited London in 1837: 'It . . . articulates with startling accuracy the words Papa, Mamma, Father, Mother, thumb, plum and some other sounds,' he noted. 'It also laughs and cries most admirably.'

Faber's talking head was more sophisticated, capable of reproducing a much greater range of sounds. The voice was activated by a bellows that set in motion an intricate artificial set of vocal cords in a chamber which could be manipulated to form a range of sounds. It was worked by something like a piano keyboard by Faber himself. He even had a clip to pinch the Turk's nose when it spoke in French. When Henry witnessed a private demonstration of Faber's talking head he wondered if the keyboard could be activated at a distance. He did not quite anticipate the telephone but, as a devout Presbyterian, thought the speaking Turk might be able to deliver Christian sermons at a distance.

Though it was an abject failure in America, Faber's talking head caught the eye of Phineas T. Barnum, the great showman who exhibited freaks and oddities around the world. One of the many hoaxes from which he had profited was the exhibition of a black slave who claimed to have nursed George Washington and sung to him as an infant. This would have made the toothless, blind, paralysed Joice Heth around 161 years old. For $1,000 Barnum bought the right to exhibit Heth and paraded her across the north-eastern states in pleasure gardens, concert halls and taverns. This was in 1835, before the abolition of slavery in the South, and Heth's appearance caused a great deal of soul-searching. Not everyone was convinced she was genuine: one suggested she was an automaton made from India rubber and whalebone by a manufacturer of gum elastic overshoes.

When Barnum learned of Faber and his automaton he promptly signed him up and in 1846 sent him to London where his agent arranged for him to put on his show at the Egyptian Hall in Piccadilly, which was renowned for its exhibitions of the weird and wonderful. The theatre impresario John Hollingshead said of poor Faber that he was 'not too clean, and his hair and beard sadly wanted the attention of a barber. I had no doubt that he slept in the same room as the figure – his scientific Frankenstein monster – and I felt the secret influence of an idea that the two were destined to live and die together. . . . There was truth, laborious invention, and good faith in every part of the melancholy room. 'As a crowning display, the head sang "God Save the Queen" which suggested, inevitably, God save the inventor.' Faber did find one distinguished admirer: the Duke of Wellington, who had a go on the automaton himself, getting it to speak German, a language with which he was familiar. He signed Faber's autograph book acknowledging his machine as a work of genius.

Hollingshead saw clearly that the unhappy German was not going to have any more success in London than anywhere else: Faber returned to America where, around 1860, he is said to have committed suicide. His automaton lived on, toured by a self-styled Professor Faber, the husband of one of his nieces. It is one of the oddest stories in the history of invention. However, Faber's miserable life was not lived in vain. One of those impressed by his demonstrations in the Egyptian Hall was Melville Bell, an elocutionist who was developing what he called Visible Speech to teach enunciation. Bell's study of the production of vowels and consonants, and of the configuration of the mouth and larynx in the creation of the range of sounds which constituted all spoken language, drew on the same intricate examination of vocal physiology that had enabled Faber to build his machine. Bell had recently married and a year after he had examined with great interest the Talking Turk his second son, Alexander Bell, was born: later, finding his name undistinguished, Alexander added the Graham.

* * *

In an article in *National Geographic Magazine* published in March 1922, the ageing, and by then world famous, inventor of the telephone recalled: 'I was . . . quite young when I had the opportunity of meeting Sir Charles Wheatstone. The interview at which I was present had nothing to do with electricity or the electric telegraph, but related to a very different subject altogether.' In the article, entitled 'Prehistoric Telephone Days', Alexander Graham Bell told how his father had taken him to see a speaking automaton that Wheatstone had constructed by following the instructions in a book written by Baron von Kempelen in the eighteenth century. Von Kempelen was most famous for the creation of a chess-playing mechanical Turk. This was a hoax, as a chess wizard was concealed beneath the chessboard and could dictate the moves of the supposed automaton from below. Von Kempelen's speaking machine, however, was a genuine all-mechanical device. It was Kempelen's book, *The Mechanism of Human Speech*, which had inspired Faber to build his much more complex machine. But Wheatstone's construction was inspiration enough for the teenage Bell.

'I saw Sir Charles manipulate the machine, and heard it speak,' Bell recalled, 'and although the articulation was disappointingly crude, it made a great impression on my mind. Sir Charles very kindly loaned my father the Baron von Kempelen's book, and I devoured it when we reached home. . . . Stimulated by my father, my brother Melville and I attempted to construct an automaton speaking machine of our own.' The brothers were encouraged by their father to put aside von Kempelen's blueprint and to create an automaton from scratch so that they would learn at first hand the wonderful mechanisms that made human speech possible. While Bell moulded the tongue and the mouth from gutta-percha, the rubber-like substance that was used as insulation in submarine telegraphy, Melville, the more skilled technician, fashioned the lungs and the throat. They decided they needed to kill a cat to examine its larynx but were too squeamish to do the deed themselves. A friend studying medicine played a cruel trick on them: saying he was putting the cat to sleep he poured acid into its mouth so that it died a horrible death. Sickened by this cruelty, they settled for a sheep's larynx provided by a butcher. Their automaton, its lungs worked with a bellows, was able to say 'Mama' convincingly enough to worry the neighbours and they left it at that.

Although it was not much more than a plaything, Bell recalled later in life that it had been an inspiration: he and his brother had learned something about the persistence required for invention and he had begun to puzzle over the nature of the human voice. A party trick that followed the automaton was his talking dog. By manipulating the mouth of the family's pet, an adopted stray Skye terrier, which he had trained to growl on cue, he got it to say 'Ow-ah-oo-gamama' which, with a bit of imagination, could be understood as 'How are you grandmamma?' The dog, keen to get its reward, attempted to speak without Bell manipulating its muzzle, but without success.

Speech and elocution were the Bell family's stock-in-trade. His grandfather, the first Alexander Bell, born in 1790, had been a shoemaker with a liking for the stage. Though his acting career was not too successful he learned much about speech while working as a prompter in the theatre, and this set him on the road to becoming an elocutionist with a special gift (so he claimed) for curing stammering. He settled in Dundee where he made a good living teaching at the Academy while giving private lessons to the children of wealthy families. His wife Elizabeth enjoyed a carriage and servants. She also enjoyed secret meetings with William Murray, rector of Dundee Academy. The lovers were indiscreet, and when Bell discovered his wife's infidelity he sued for divorce and tried, unsuccessfully, to sue Murray as well. This scandal wrecked his career in Dundee. The family broke up and he took his son, Alexander Melville, to London.

The histories of every scientific advance, and every innovation, involve elements of chance and it is sometimes asked if the world might be missing some of its familiar technologies – television, the aeroplane, or the telephone – if fate had not directed the path taken by the inventor. John Baird's illness took him to recuperate in Broadstairs, where he found inspiration and practical assistance in his quest to 'see with electricity'. If Wilbur Wright had not suffered injury as a teenager it is unlikely that he would have abandoned his studies to mend bicycles and fly aeroplanes. Fate undoubtedly plays a part in guiding an individual towards an invention, and this was more than usually true of Alexander Graham Bell and the telephone. But this does not mean, of course, that without his inspiration there would be no telephone. As with Baird's first television, the Wright Brothers' first Flyer and Bell's first, crude, telephone the time was clearly ripe for the breakthrough and there were always others seeking the same 'eureka moment'. The answer to the old chestnut: 'Would we have the light bulb if Thomas Edison had not been born?' is an unequivocal 'Yes!'

The same is true of the telephone: others were close to making the breakthrough when Bell lodged his first patent. However, the story of how he came to get there first is especially interesting as he was up against at least one rival who was already an established and successful inventor. Why, it has been asked by David Hounshell, Professor of Technology and Social Change at Carnegie Mellon University, did Bell, a rank amateur as he freely admitted, with no knowledge of electricity, get there first? Was it because he was a novice that he was able to imagine that a 'speaking telegraph' might prove to be popular while the 'professionals', engrossed in the problems of the existing telegraph system, could see no market for it? As it turned out, not everyone was any more impressed with the first telephones than they were with the hapless Faber's talking automaton. Notoriously, the chief engineer of the British Post Office, Sir William Preece, told a Parliamentary Committee that he thought the telephone more suited to America than to Britain, where there were sufficient telegram boys to keep everyone well informed.

The first step on Bell's road to his 'eureka moment' was taken before he was born. His grandfather remarried after his move to London, published a book, *The Practical Elocutionist*, and had regained his reputation as a celebrated 'Professor of Elocution'. However his son, Melville, who had moved to London with him, was not prospering. He was an assistant in a draper's shop and clearly unhappy. Concerned about his health, in 1838 his father arranged for him to stay in St John's, Newfoundland with a family friend. His job as a clerk in a shipping office was not exciting, but he thrived in St John's, taking part in amateur dramatics, giving classes on Shakespeare and offering some of the expertise on elocution he had picked up from his father. When, after four years, he came back to London he was ready to set up as an elocutionist in his own right and he held an abiding affection for North America. In 1844 he married Eliza Symonds, who was nearly ten years older than him, a spinster living in an apartment in Edinburgh with her widowed mother. Eliza was deaf and could only hear with an ear trumpet. It was not, as Melville wrote later, 'love at first sight', but the marriage was successful and the couple settled in Edinburgh.

Three sons were born in as many years: Melville, known as Melly, in 1845 Alexander, known as Aleck, in 1847 and Edward, known as Ted, in 1848. They were brought up in Edinburgh, schooled at home by their mother at first, and then in various colleges. Despite her deafness, Eliza was an accomplished pianist and for a time Aleck had the ambition to be a musician. He showed little aptitude for school work. When he was thirteen he spent a year in London with his grandfather; this, he said later, turned him into a studious young man. It was in that year that he visited Wheatstone with his father.

The boys were very much in thrall to their father whose great project was 'Visible Speech'. The sounds made by the human voice were analysed and then rendered as a set of symbols on a chart devised by Melville. He taught the boys to recognise each of these symbols and to make the sound they represented. A word would be made up of several symbols, which he could write out. To illustrate how it worked he would ask an audience to speak words in any language, whether he understood it or not. It might be Greek or Arabic. He would chalk up the symbols representing the sounds of the words and then call in Aleck and Melly to interpret them. The results were impressive. Though they might have not the slightest idea what they were saying they would pronounce a word correctly: anticipating, perhaps, Wittgenstein's 'If a lion could speak we would not understand him.'

One practical application of Visible Speech was as an aid in teaching the deaf to speak and Aleck became proficient at it. To the end of his life he liked to describe himself as a teacher of the deaf. However, it was not at all clear what his professional future might have been, had tragedy not struck his family. First his younger brother Ted fell ill, dying of tuberculosis in May 1867 at the age of eighteen. Three years later his older brother Melly, who was married and had had a child who had died, also succumbed to tuberculosis. That was at the end of May 1870 and it determined Aleck's future in a way he could not have anticipated. His father had visited Canada and America as a lecturer since his first experience of the New World in Newfoundland. It seemed to him and his wife that for the health of their surviving son they should emigrate. Aleck took some persuading, protesting that he did not want to be dragged into 'the backwoods'.

They set sail on 21 July 1870, less than two months after Melly's death, arriving in Quebec on 1 August before the onward journey by steamer to Montreal. For the trip Aleck had a French translation of a book by the German physicist Hermann von Helmholtz entitled *On the Sensations of Tone*, which, in time, was to influence his thoughts on telephony. The Bells visited friends who had settled in Paris, Ontario and found a place for themselves near the town of Brantford. For $2,600 they were able to buy a small estate in a place called Tutelo Heights: ten-and-a-half acres with orchards, a variety of outbuildings, stables and animal pens, as well as an icehouse in which to store a winter harvest. The house itself was a good size, with four bedrooms and a conservatory. It was here that Aleck began to recuperate, for he too felt that he might be suffering from a terminal illness.

He was not 'in the backwoods' for long. When his father was invited to teach Visible Speech at Boston School for Deaf Mutes he suggested his son could go in his place. Aleck took up the appointment in March 1871 and was introduced for the first time to the lively intellectual world of Boston where his father had friends and admirers. After a summer back in Brantford he returned to Boston that autumn and began to take on private pupils. For the next four years Bell pursued his career as a teacher of the deaf and was rewarded in 1873 with a professorship in Vocal Physiology and Elocution at Boston University. In private, and with great secrecy – he had a special table with a lockable top for his experiments – he began to dabble in electrical experiments. His first idea was that his knowledge of the different timbres of the human voice might somehow lead him to offer a solution to a problem which was taxing the rapidly expanding electric telegraph companies. Demand was outstripping the capacity of the system. Thomas Edison had invented a quadruplex telegraph system. Bell thought he might be able to double that.

However, while pondering the difficulties of creating a multiple telegraph system, the possibility that the vibrations produced by speech might activate some kind of electrical impulse which could be sent along a telegraph line and then transformed back into speech began to dominate his thoughts. Faber's automaton, which turned up again in Boston worked by its new owner 'Professor Faber', was one inspiration. Another was a French invention, the phonautograph, in which spoken words were recorded as patterns on paper. With the help of a friend who was a surgeon, Bell built his own version of this device using parts of the ear of a cadaver as a mechanism which, sensitive to his voice, moved a straw stylus on a piece of paper. And there was the manometric flame in which changes in voice tone altered the shape of a gas flame. The idea that a voice might provide sufficient power to generate an electric current was taking shape. However, like all inventors, Bell needed financial backing. He had already received some from the wealthy father of a boy he taught. It was another pupil who put him in touch with money, enthusiasm and expertise and a great deal else besides.

In the autumn of 1873 the fifteen-year-old daughter of a successful lawyer, Gardner Greene Hubbard, became one of Bell's pupils. Scarlet fever had destroyed Mabel Hubbard's hearing when she was five years old. Before she became deaf she had been able to speak but the danger was that she would gradually lose the ability. Her parents were determined that she should not be reduced to communicating silently with sign language and made every effort to encourage her speech. Bell was the ideal tutor and he and Mabel got on well. When he was experimenting both with the multiple telegraph idea and the 'speaking telegraph' he asked Gardner Hubbard if he would help to finance his research. This Hubbard was only too keen to do and, as a patent lawyer, immediately began to investigate the legal and commercial possibilities. The multiple telegraph looked like a good bet. Bell illustrated one of his ideas about the way in which sound vibrations could produce a response in wire by humming into a piano at different pitches.

Hubbard was impressed, but warned Bell that he was in a race to produce a multiplex telegraph. On 23 November 1874 Bell wrote home to his parents: 'It is a neck and neck race between Mr Gray and myself who shall complete our apparatus first. He has the advantage over me as a practical electrician – but I have reason to believe that I am better acquainted with the phenomena of sound than he is – so that I have the advantage there.' What he was not aware of was the fact that this same rival was close to creating not only a multiplex telegraph but a speaking telegraph, his inspiration, remarkably, a popular plaything nicknamed the Lovers' Telegraph.

* * *

In volume three of his monumental work, *Nature's Miracles*, published in 1900, one of the foremost American telegraph engineers, Elisha Gray, told the following story of how he nearly invented the telephone in

1875 but was discouraged by his financial backers. Gray had by turns amazed and alarmed the community of telegraph operators by transmitting electronically generated popular tunes on parts of the network and his mind had turned to transmitting speech. He was in Milwaukee in Wisconsin conducting some experiments when he had a revelation:

> One day while out on an errand I noticed two boys with fruit-cans in their hands having a thread attached to the center of the bottom of each can and stretched across the street, perhaps 100 feet apart. They were talking to each other, the one holding his mouth to his can and the other his ear. At that time I had not heard of the 'lovers' telegraph', although it was old. It is said to have been used in China 2000 years ago.
>
> The two boys seemed to be conversing in a low tone with each other and my interest was immediately aroused. I took the can out of one of the boys' hands (rather rudely as I remember now), and putting my ear to the mouth of it I could hear the voice of the boy across the street. I conversed with him a moment . . . when, suddenly, the problem of electrical speech-transmission was solved in my mind.

Preoccupied with his work on the multiplex harmonic telegraph for which he had considerable financial backing, he put the idea for the telephone to one side. He did, however, mention it casually to his partner. 'He gave me a look that I shall never forget, but he did not say a word. The look conveyed more meaning than all the words he could have said, and I did not broach the subject again.'

Elisha Gray was a child of the telegraph era. He was born in 1835 the son of a farming family in Barnesville, Ohio. When he was just ten years old he built his first telegraph line. He had to leave school two years later after the death of his father and began work as a carpenter and boatbuilder. In his spare time he took an interest in science and kept up his boyhood interest in electricity. When he was in his twenties he was able to study at Oberlin College, which had been founded in 1833 by a pastor and a missionary with progressive views: they admitted black Americans and the college became a stop on the underground for escaping slaves. Studying and working took a toll on his health, but after marriage at the age of thirty he began to gain a reputation as a telegraph inventor. His first patent, in 1867, for a telegraphic relay, caught the attention of the Western Union Telegraph Company. He was loaned money to buy an interest in a telegraph instrument shop in Cleveland, Ohio. With a partner he moved to Chicago to set up a manufacturing company. By 1875 the company was profitable enough for him to sell up and to become an inventor full time.

Gray's experimentation with the relationship between electric frequencies and sound was encouraged by a chance discovery made by a young nephew who was staying with him. For his amusement the boy had wired himself up to a zinc bathtub so that he became part of a low-level electrical circuit. He discovered that when he touched the tub a tone was created. Gray conducted a number of experiments and discovered be could vary the tone by the way in which has hand made contact with the bath. This was a discovery that could, with considerable development, be incorporated into the creation of the holy grail of telegraphy: a multiplex transmission system that could handle eight signals simultaneously.

So in 1875 Gray and Bell were working on the same potentially very valuable invention. Both their backers, Hubbard in Bell's case, and Western Union in Gray's, wanted them to concentrate all their efforts on that and not to be distracted by any thoughts about a 'speaking telegraph'. The showcase for the multiplex systems would be the Centennial Exhibition in Philadelphia in 1876. Gray's exhibit would include an entircly new telegraph line following the Pennsylvania Railroad from Philadelphia to New York on which he would show how he could multiply the calls on a single line.

He was aware of Bell's plans to add the speaking telegraph to his exhibits in Philadelphia but was not unduly concerned. He had written to his patent lawyer in October 1875: 'Bell seems to be spending all his energies in the talking telegraph. While this is very interesting scientifically it has no commercial value at present, for they can do more business over a line by methods already in use than by that system.'

Although Wheatstone in England had sold personal telegraph lines which required no trained operators, in America the notion that you could do away with an army of Morse telegraphers and telegram boys and enable people to talk to each other directly seemed fanciful. The telephone, if it were made to work, would be nothing more than a toy, of no greater significance than the Lovers' Telegraph. Perhaps Bell would not have continued to experiment with it if his version of the multiplex telegraph had he not been rudely rebuffed by Western Union who had their own man Elisha Gray. But there was another positive influence too: none other than the grand old man of American electrical experimentation Joseph Henry, secretary of the Smithsonian Institution in Washington.

* * *

Just as Morse needed Alfred Vail as his technician so Bell needed his. He found him in the Boston machine shop of Charles Williams, which served from time to time as his laboratory. Thomas Watson was the son of a livery stableman in Salem, just to the north of Boston, where Bell was living at the time. He left school at the age of fourteen keen to find work. He tried bookkeeping and then carpentry before he found a job to his liking in Charles Williams's workshop in 1872. In 1874 he was assigned to work with Bell on his multiple telegraph and began to make equipment to Bell's specifications. The two got on well, Bell impressed with Watson's skills as an instrument maker, and Watson struck by Bell's erudition and refined manners. There was a camaraderie that was absent from Morse's relationship with Vail who, for long periods, had been left alone to get on with refining the telegraph. Watson's first task was to make sets of vibrating reeds, which Bell hoped would form the basis of his multiplex telegraph.

Watson proved to be a willing and able instrument maker but Bell felt he was still lacking in electrical knowledge and in March 1875 decided to consult the seventy-nine-year-old Joseph Henry. He was received cordially, if without great enthusiasm at first. Bell described what he was trying to achieve with his harmonic telegraph and told Henry of the sounds that would emanate from an empty coil. Henry asked to see Bell's equipment and for permission to repeat his experiments and publish the results. He was so enthused that he offered, despite his age, to come to see Bell but he was saved the trouble. Bell took his equipment to the Smithsonian. It was while he was demonstrating the effects he could achieve that he broached the subject of the telephone. Henry seemed to have forgotten that it was an idea he had had thirty years earlier when he witnessed a performance of Faber's automaton, for he reacted to Bell's idea as if it were a novelty. This was the spur Bell needed and he wrote home:

> I explained the idea and said 'What would you advise me to do – publish it and let others work it out – or attempt to solve the problem myself?' He said he thought it was 'the germ of a great invention' and advised me to work at it myself instead of publishing. I said I recognised the fact that there were mechanical difficulties in the way that rendered the plan impractical at the present time. I added that I felt that I had not the electrical knowledge necessary to overcome the difficulties. His laconic answer was – GET IT. I cannot tell you how much those two words have encouraged me.

In the end, it was a chance observation that gave Bell a clue as to how to fulfil his speaking-telegraph ambition. As Watson remembered it, the revelation came on 2 June 1875 when the telegraph system they were trying to perfect was playing up. A vibrating reed had stuck and he made several attempts to get it going by giving it a twang. Bell suddenly realised that he had heard the sound it made transmitted along the wire. He asked Watson to make the first ever model of a telephone immediately, a task he had completed by the next day. It was crude, involving a vibrating diaphragm activated by a steel reed, but it worked well enough for Watson to make out Bell's voice, though what he said was unintelligible.

It was another nine months before Watson and Bell had modified the system so that the speaking telegraph was able to transmit clearly articulated words. On 10 March 1876, in a laboratory Bell had hired away from the noise of the Williams machine shop, Watson clearly heard his name spoken in the receiver: 'Mr Watson, come here, I want to see you.' Watson, in turn, read passages from a book for Bell to hear. The sound quality was still poor but he was confident that he had in his hands a brilliant invention. He wrote to his father that evening: 'I feel I have at last struck the solution of a great problem – and the day is coming when the telegraph wires will be laid on to houses just like water and gas – and friends converse with each other without leaving home.'

Despite his apparent confidence in his invention, Bell was reluctant to demonstrate it personally at the Philadelphia Centennial Exhibition in 1876. He was persuaded to go by Mabel Hubbard, with whom he was now madly in love and whom he was planning to marry. Her father was urging him to go too, for fear of Elisha Gray getting all the attention with his multiplex telegraph. Bell had several exhibits: Visible Speech, his version of the harmonic telegraph, what he called a teleautograph, which transmitted handwriting and, tucked away at one of end of the huge exhibition halls, his telephone. In one of the more bizarre turns of fate in the Bell story his first publicity coup was launched by the portly and vastly bearded Emperor Dom Pedro of Brazil. Bell had invited this dignitary to witness a demonstration of Visible Speech in Boston and it had been a great success. Accordingly, in Philadelphia, Dom Pedro was keen to see Bell's telephone. In his excitement at hearing words spoken in the receiver he exclaimed: 'I hear, I hear.' Only a handful of scientists witnessed the demonstration in which an extract of Hamlet's 'To be, or not to be' soliloquy was heard. One of those who listened in was Elisha Gray. He could not make out the words at first and then announced that he had distinctly heard: 'Aye, there's the rub.'

The excitement in Philadelphia was momentary. There was only one newspaper report about it so there was an urgency now for Bell and his backers to advertise the telephone. Bell, aided by Watson bellowing songs into the receiver, gave a series of public lectures, which in time interested the newspapers and scientific magazines. In the spring of 1877 they made their first long-distance phone call from Boston to Cambridge, Massachusetts.

On 11 July, with the Bell Telephone Company founded and beginning commercial operations, Mabel and Alec (she persuaded him to drop the k) were married in the Hubbards' Cambridge home and set off for their honeymoon, going first to Niagara Falls and then on to Brantford, where Mabel met Alec's mother for the first time. At the party held to celebrate the wedding it was arranged for Chief Johnson, head of the native Six Nations, to talk in Mohawk on a telephone line laid between the Bells' family home in Tutelo Heights and Brantford. In August the couple sailed from New York for Plymouth, leaving Watson, now a shareholder of the Bell Telephone Company, to carry on with improvements to the equipment. Bell himself, astonishingly, had left London only seven years earlier heading for what he feared would be the 'backwoods'.

When the Western Union Telegraph Company realised the great commercial potential of the telephone – Bell's patent is often described as the most lucrative ever granted – they made a last ditch attempt to claim precedence for Gray. The court battles dragged on but Bell's patent stood the test. Watson worked for Bell for seven years and then left the Bell Company in which he had a lucrative stake. Later in life he took up acting, with a special liking for Shakespeare, which is perhaps why, in a recorded interview he gave towards the end of his life, he has no trace of an American accent.

Bell became immensely wealthy: in 1879 the value of shares in the telephone company rose from $65 to $1,000 in a matter of months. He and Mabel were generous with their wealth. When Professor Faber, with his inherited automaton, appealed for financial help Bell gave him $500, recognition for one of the inspirations that led to the creation of the telephone and his fortune. Yet while telephone technology rapidly improved Bell played no further part in its development, turning his mind to other interests, notably the problem of heavier-than-air flight.

However, one project involving the telephone he regarded as his most brilliant invention: what he called his 'photophone'.

The inspiration for this was the discovery of the photoelectric properties of the element selenium, the same element that was used by John Logie Baird in his pioneer attempts to 'see by electricity'. With a newly acquired assistant, twenty-five-year-old Charles Sumner Tainter, Bell devised a telephone system in which sound was transformed into light and back again into sound. It had its limitations, for it worked only in line of sight and with sufficient light. However, he wrote in great excitement to his father:

> Can imagination picture what the future of this invention is to be! . . . We may talk by light to any visible distance without any conducting wire. . . . In warfare the electric communications of an army could neither be cut nor tapped. On the ocean communication may be carried on . . . between vessels.

With Tainter's help, Bell managed to make a working photophone, an invention that went from eureka to obsolescence in a matter of a few years. Bell had the right idea but, this time, he was unaware of the scientific breakthrough that would, in a short time, make wireless telephony a reality.

* * *

It was August 1898 and the Cowes Regatta on the Isle of Wight was in full swing, with a great deal of nobility and royalty afloat. On the Royal Yacht *Osborne* Queen Victoria's son, Prince Edward, known affectionately as Bertie, was nursing a knee injury sustained at a ball in Paris in one of the Rothschilds' palaces. Her Majesty was staying at her favourite summer residence, Osborne House on the Isle of Wight, and was anxious to get news of Bertie's recovery. The Royal Yacht, however, was not at anchor but was cruising in the Solent and along the south coast. How could she get news of that damaged knee? The Queen was seventy-nine years old but had no apparent fear of new technology. She summoned the young man who had recently used his new wireless technology to report on the Kingston Regatta in Ireland to send messages from ship to shore. Mr Marconi, an Italian with a rather refined English accent, was only too happy to oblige. The year before he had established the world's first wireless telegraphy station at the Royal Needles Hotel at Alum Bay on the Isle of Wight for his exploration of signalling from ship to shore. The pleasure boats that sailed from pier to pier along the south coast in the summer were ideal for his experiments with mobile messaging.

While assistants raised an 83-foot aerial on the Royal Yacht, Marconi himself set up a station in Ladywood Cottage in the grounds of Osborne House equipped with an aerial 100 feet high. All of this was put together with impressive speed. There was very little in the way of cable to be laid. Messages were to be sent miraculously through 'the ether'. They could not be spoken: the technology could not handle that – but it could transmit in Morse Code with a telegraph operator listening in and jotting down the letters of the dots and dashes.

It was in this way that a message from the Royal Yacht *Osborne* was sent to the Queen: 'H.R.H. the Prince of Wales has passed another excellent night, and is in very good spirits and health. The knee is most satisfactory.' After the first message was sent on 4 August those aboard the *Osborne* took great delight in making use of the new technology. One message read: 'Very anxious to have cricket match between *Crescent* and Royal Yachts Officers. Please ask the Queen whether she would allow match to be played at Osborne. *Crescent* goes to Plymouth Monday.' From Ladywood Cottage the reply was tapped out: 'The Queen approves of the match between the *Crescent* and Royal Yachts Officers being played at Osborne.' It was such fun. Emily Amptill at Osborne House asked Miss Knollys on the yacht: 'Can you come to tea with us some day?' Miss Knollys replied: 'Very sorry cannot come to tea. Am leaving Cowes tonight.' Around 160 messages were sent in all. The Prince of Wales presented Marconi with a fabulous tie-pin and the Queen granted him an audience. But the brush with royalty was not what excited Marconi most during the days he spent keeping the *Osborne* in touch with the mainland. He was still not sure how far his wireless waves would travel and was intrigued to discover that wherever the Royal Yacht went he seemed to be able to keep in touch with it.

It was a complete mystery to those who saw wireless demonstrated how this twenty-four-old young man had chanced upon his miraculous invention. And he could not quite explain how it worked himself. The telegraph and the telephone, now commonplace, had been extraordinary enough when they first appeared. But at least you could see that something was happening with wires. With Marconi's wireless the signals that carried the dots and dashes of Morse Code message were themselves invisible and inaudible – and they appeared to penetrate solid walls and hills and to keep track of a moving ship. Something of the mystery surrounding wireless and its inventor was captured by a reporter for the *Daily Express* in Dublin who had watched Marconi when he covered the Kingston Regatta using equipment mounted on an old steamship the *Flying Huntress*. He sent back commentary to a station in the harbourmaster's house:

> A tall, athletic figure, dark hair, steady grey blue eyes, a resolute mouth and an open forehead – such is the young Italian inventor. His manner is at once unassuming and yet confident. He speaks freely and fully, and quite frankly defines the limits of his own and of all scientists' knowledge as to the mysterious powers of electricity and ether. . . . Signor Marconi listens to the crack-crack of his instrument with some such wondering interest as Aladdin must have displayed on first hearing the voice of the Genius who had been called up by the friction of his lamp.

With Marconi on the *Flying Huntress* was his loyal and indefatigable assistant George Kemp, who had joined the British Post Office after service in the Royal Navy. A stocky chap with a handlebar moustache, he knew how to handle masts and ropes. Interviewed by the *Express* he made it clear he had no time for theories and was disdainful of scientific discovery: 'The one thing to do if you expect to find out anything about electricity is to work, for you can do nothing with theories. Signor Marconi's discoveries prove that the professors are all wrong, and now they will have to go and burn their books.'

In his blunt and ill-informed way, Kemp was echoing the opinion of Sir William Preece, his former boss at the British Post Office. For a number of years Preece had been locked in bitter argument with a group of scientists and mathematicians who had acquired the sobriquet the Maxwellians. They were believers in the theoretical work of a Scottish physicist, James Clerk Maxwell, who had studied the experimental discoveries of Faraday and the observations about the behaviour of electricity afforded by work on the telegraph and had arrived at the conclusion that the invisible forces created by electric currents moved in waves. These waves were essentially the same as light waves but they were a different length, as measured by the distance between the crests of the waves. In 1865 he published *A Dynamical Theory of the Electromagnetic Field*, which was to have a huge influence on subsequent research and speculation both in Britain and Europe. It was couched in the form of mathematical equations and was therefore accessible only to those with a similar grasp of the logic.

Maxwell, who was a Scot from a wealthy landowning family, died in 1879 but fascination with his work continued and his theories were refined over a number of years. It was not clear that they would have any practical value and it was more than thirty years after the publication of his electromagnetic theory that it gave rise to the wireless telegraphy that Marconi demonstrated to Queen Victoria in 1898. Marconi, like Kemp and Preece, were what one of the Maxwellians called 'practicians': they had little or no scientific understanding of electromagnetism but worked by trial and error. Early on Marconi liked to describe himself as an 'ardent amateur student of electricity'.

The interplay of theory and practice which created wireless telegraphy is one of the strangest tales in the history of invention and gave rise to a dispute about who was first in the field which continues to the present day. Was it the leading Maxwellian Professor of Physics and Mathematics at University College Liverpool, the towering and impressive Sir Oliver Lodge, or was it the suave young Italian Guglielmo Marconi? It was the death in 1894 of the brilliant German physicist and experimenter Heinrich Hertz at the age of thirty-six that brought Lodge and Marconi into conflict. Whereas Lodge had known Hertz personally and had conducted very similar laboratory experiments in the 1880s, Marconi learned about the German's great discovery from an obituary in an Italian electrical journal. The author of the piece was Augusto Righi, a professor at Bologna University whose lectures Marconi had sometimes attended. Hertz had died in January and Marconi did not read about his experiments until the summer when he was on holiday in the Italian Alps with his older half-brother Alfonso. He cared nothing then about the Maxwellians or the history of electromagnetics. What fired his imagination was the conviction that an experiment Hertz had conducted involved sending an electronic signal across a small space in the laboratory to activate a 'receiver'.

Hertz had found that he could generate and detect at a distance an electric charge that travelled through space in waves the length of which he could measure. It did not occur to him that it might be the basis of a new form of communication. But this did occur to Marconi. 'My chief trouble was that the idea was so elementary, so simple in logic that it seemed difficult to believe no one else had thought of putting it in practice,' his eldest daughter Degna recalls him saying in her book *My Father, Marconi*. In fact Oliver Lodge had, but he had missed the correct answer by a fraction. 'The idea was so real to me that I did not realize that to others the theory might appear quite fantastic.'

Lodge had given the Hertz Memorial Lecture at the Royal Institution on 1 June 1894 and had replicated some of the German physicist's laboratory experiments demonstrating the transmission of electromagnetic waves. The instruments used for this are not easy to describe in lay terms and it is not quite clear what they were in actuality. However, as Marconi acknowledged, Lodge and he had begun working on the same lines but Lodge had not proposed adapting the equipment for use in telegraphy – certainly not in the Royal Institution lecture. In August Lodge gave two more lectures in Oxford, demonstrating Hertzian electromagnetic waves: one was entitled 'Experiments Illustrating Clerk Maxwell's Theory of Light', the other 'An Electrical Theory of Vision'. It was in these lectures that Lodge claimed that he had used a Morse key and had demonstrated the sending of a message; this, if true, would have given him precedence in the invention of wireless telegraphy.

Guglielmo Marconi was the archetypical amateur inventor in the sense that he was largely self-taught. But he never had to struggle to find funding and in that he was exceptional. His mother was Annie Jameson, from the Irish whiskey family, his father Giuseppe Marconi, a moderately wealthy landowner with an estate at Pontecchio in the countryside a few miles from Bologna. They met when she was sent by her family to Italy to learn the bel canto style of operatic singing. Giuseppe was a widower with a young son, Alfonso. They fell in love but the Jameson family forbade her to marry Giuseppe. Annie defied her family, kept in touch with him secretly, and they eloped to the French seaside town of Wimereux where they married. They lived in his home, the Villa Griffone in Pontecchio, and a town house in Bologna.

Marconi was born in Bologna in 1874. He grew up with the freedom to roam the Villa Griffone estate with Alfonso. From an early age he showed signs of wanting to invent something, but he did not know what. He experimented with electricity and his father gave him the money to buy batteries and other equipment and a subscription to technical magazines. His mother made sure he spoke English and she frequently took the boys to stay with relatives at Leghorn. It was there that Marconi as a teenager was taught to use a Morse key by an elderly and blind telegrapher in return for reading him stories. He enjoyed sailing and had at one time thought of joining the Italian Navy but he did not gain the relevant qualifications. Meanwhile his mother indulged his amateurish experiments.

When, at the age of twenty, Marconi had his inspiration about the use of Hertzian waves for telegraphy his mother made available to him an area in the attic of the Villa Griffone, which had been used for raising silkworms. It was in this domestic workshop that he set about learning everything he could about how electromagnetic waves were sent and received. Sending was relatively simple: you created a spark by passing a current, amplified by a coil, across two brass balls. The best available blueprint for a receiver was something that became known as a 'coherer'. The observation that metal filings would cling together when hit by an electromagnetic wave was the inspiration for this clever device, which had been developed by a French physicist Edouard Branly and adapted by Lodge to create an on–off switch that could be operated at a distance. Metal filings were placed in a sealed glass tube with a wire connection at either end and coupled to an electrical circuit. When the filings were loosely scattered the flow of electricity was negligible and the circuit was 'off' but when the filings were struck by an electromagnetic wave they stuck together (cohered), closing the circuit and turning it 'on'. To turn the switch off again the filings had to be scattered.

Marconi did not invent this mechanism but he greatly refined it for use in telegraphy, working long hours. He adapted thermometers for the coherer case and experimented to find the most sensitive metals for the filings. A brilliant instrument maker, he was able to devise a little hammer mechanism which tapped the tube to break the circuit so that it could be turned on and off automatically. Morse Code was a gift: it meant that all Marconi had to do was find a way to get his receiver to register long and short pulses.

The great unknown was how far electromagnetic waves could be sent. Because it was by then accepted that they were the same as light waves, and a light beam would not follow the curvature of the earth, their range would be limited: they would just fly off into space. However, Hertzian waves were longer than light waves and some conjectured that this might make a difference. Marconi had no clue at all what would happen when he first moved out of his attic laboratory into the fields around the Villa Griffone. To test the distance the waves would travel he did not need at first to send messages: a bell attached to an electric circuit which was closed when hit by a wave was sufficient. Alfonso, and sometimes one of the estate workers, would indicate whether he had received a signal: when they were in sight it was with a wave of a handkerchief; when they were on the other side of the hill behind the villa, a shot from a hunting rifle.

Although others were experimenting with Hertzian waves, nobody was working as strenuously as Marconi to devise a system for practical, commercial wireless telegraphy. He worked on a theory that the longer the waves the further they would travel. To increase the range of his signals he earthed the transmitter and rigged it up to a wire aerial. He really had no theory to help him. Very little was known about the potential of these electromagnetic waves. If there was received wisdom it was that, for one reason or another, they might not get very far.

Marconi's mother had always encouraged him while his father had been dubious about the invention. However when Giuseppe realised that his son was sending signals more than a mile and over a hill he backed his son. There is a family story that they made the first offer to the Italian Navy, which turned it down. No record of this has been found and very soon Marconi and his mother were off to London where family connections and British maritime dominance held out greater promise than anything in Italy. Early on Marconi had recognised that wireless telegraphy would bridge a gap in the already extensive international cable network: communication between ship and shore, and ship to ship. There was also a possibility that wireless would be cheaper than wire telegraphy as there was no need to lay cables.

In London, a relative of Marconi, Henry Jameson Davis, who had known him as a boy, gave him the introduction which led him to William Preece, chief engineer of the all-powerful General Post Office. Preece immediately recognised a kindred spirit, not a Maxwellian but a 'practician'. He arranged for Marconi to demonstrate his still primitive equipment and promised government backing. Preece himself had already attempted wireless telegraphy with some success. This exploited induction – the fact that electricity would jump from one wire running parallel to another. The distance the signal would travel was related to the length of the wire. Preece used it to bridge channels between the coast and small islands. However, when he tried jumping a signal across the Irish Sea from the west coast of England to the east coast of Ireland it failed: the range was strictly limited.

It seems Preece did not understand the fundamental difference between wireless induction and the sending of electromagnetic waves. Marconi himself, in interviews, said he did not know if his signals were carried on Hertzian waves: perhaps he had by chance discovered something new. It really did not matter to him one way or the other: he was intent on discovering how far they could travel. Distance was vital if his telegraphy was to be commercial. The only way he knew to extend the range of his signals was to raise larger aerials and generate more powerful sparks. This he did methodically, his progress chronicled by newspapers and magazines.

Understandably, Lodge and the Maxwellians were furious with Preece for parading Marconi before the public. The Irish professor George Fitzgerald wrote in 1896 to the eccentric genius, and nephew of Charles Wheatstone, Oliver Heaviside:

> Preece surprised us all by saying that he had taken up an Italian adventurer who had done no more than Lodge and others had done in observing Hertzian radiations at a distance. Many of us were very indignant at this over-looking of British work for an Italian manufacturer. Science 'made in Germany' we are accustomed to but 'made in Italy' by an unknown firm was too bad.

In fact, Marconi did not stay long with Preece. When the promised government funds were not forthcoming Marconi was persuaded to embark on his first commercial venture. His father sent him £300 to cover legal fees, and Jameson Davis raised £100,000 from venture capitalists to fund the Wireless Telegraph and Signal Company, which opened its offices in Mark Lane in the City of London in July 1897. This company bought Marconi's British patents as soon as they were granted and he was off to the Isle of Wight to set up his first permanent station at the Royal Needles Hotel. From then on his progress was astonishingly rapid as he demonstrated that, whatever waves he was using, he could send signals hundreds of miles and they would somehow travel through or over or under hills.

In March 1899, with his financial backers urging him to take his invention to America, Marconi chose to send a signal across the English Channel, setting up his French station in the little resort town of Wimereux where his parents had married secretly thirty-five years earlier. The American magazine *McClure's* was there to witness this landmark, and to make sure they were not being hoodwinked reporters on either side of the Channel exchanged messages in their own cipher which allowed for no trickery. The reporter concluded, 'here was something come into the world to stay'. It lasted, in fact, for a century: wireless Morse Code was not abandoned for long-distance maritime communication until 1999.

Marconi's ambitious programme of extending the verifiable range of his signals culminated in his announcement on 12 December 1901 that he and Kemp had distinctly heard in St John's, Newfoundland the three dots of the Morse 's' in a phone receiver sent from a giant spark transmitter at Poldhu in Cornwall, a distance of 1,600 miles. It was a triumph and Marconi and Kemp were fêted at the annual dinner of the American Institute of Electrical Engineers in New York on 13 January 1902. Alexander Graham Bell sent a telegram, which read: 'Hearty congratulations, I rejoice in your success.' Plaudits were read out at the dinner from, among many others, Thomas Edison and Nikola Tesla. Marconi and Kemp had received those three dots with an aerial attached to a balloon. The project cost his company £50,000, much of it spent on the Poldhu transmitting station designed with the expert assistance of Ambrose Fleming, professor of Electrical Engineering at University College London who was hired as a consultant.

In the early 1900s many of the large Atlantic liners were equipped with Marconi wireless. None could transmit right across the Atlantic but they could relay messages to each other so that at any one time the whole maritime community was potentially in touch. However, only the largest liners had two Marconi telegraph operators working in shifts:

those with just one operator had to go off air when he went off to sleep (the staff were exclusively young men). The heroics of the Marconi telegraphers in the rescue of passengers on stricken ships were already legendary when, in the early hours of 15 April 1912, Jack Philips on the *Titanic* tapped out in company jargon to a fellow operator: 'It's a CQD old man, we have hit a berg and we are sinking.' CQ was the routine signing-on call (seek you) and the D was for distress. SOS was just becoming the standard emergency code and Philips used that as well. He perished, but his junior, Harold Bride, lived to enjoy a hero's welcome in New York. Marconi became an international hero: though 1,496 perished, his wonderful invention had saved 712 lives.

But already, in 1912, there were problems with the rapid spread of wireless telegraphy. Amateur wireless hams had set up all along the East Coast of the United States where there was no regulation of wavelengths or transmissions. There was a tremendous amount of interference and schoolboys would routinely jam the experimental US Navy transmissions for fun. It came to a head with confusion over messages received on the night of the *Titanic* disaster. Reports came in that everyone was saved and the stricken liner was being towed into Halifax. The garbled messages, so distressing when they proved to be false, were blamed on the amateurs. The US Government legislated, confining them to a wavelength of 200 metres or less, believing, wrongly, that this would limit their range. In 1914 the American Radio Relay League was formed and the amateurs promptly established their own continent-wide network of stations. The frequencies they used were regulated and very soon it became clear that wireless was going to become as congested as the wire telegraph had been when only two messages at a time could be sent. There was a great deal to learn about the frequencies that the 'ether' had to offer, in particular how to prevent one signal interfering with another.

The most puzzling question was how it was that Marconi's waves, which were, in fact the same electromagnetic pulses first discovered in the laboratory by Hertz, could travel more than a thousand miles. One of those who guessed correctly was Oliver Heaviside, the staunch

Maxwellian: signals were being reflected from particles in the upper atmosphere, now called the ionosphere but once known as 'the Heaviside layer'. Oliver Lodge did not take a great deal of interest in this unravelling of the nature of Hertzian waves as he became enthralled by psychic research. He was a believer in the afterlife, as were many scientist contemporaries of his. In 1927 he persuaded John Reith, the first director general of the BBC, to allow him to carry out an experiment in telepathy on the radio. Members of the Society for Psychical Research attempted to 'transmit' images to listeners all over the world: a bunch of white lilac, a man in a mask, a Japanese print and two different playing cards. Twenty-five thousand impressions were received by the Society but few bore any resemblance to the original. Lodge died on 22 August 1940 believing to the end that there were spirits 'on the other side'.

Though he shared the 1909 Nobel Prize in Physics with the German scientist Ferdinand Braun, Marconi continued to take little interest in theory. But he went on experimenting and discovered that short-wave wireless signals could travel long distances. However, he does not appear to have given wireless telephony a thought. Marconi was enthralled with telegraphy, his companies had been established all over the world, and he had no incentive commercially to experiment with the transmission of speech. In time his company would be involved in radio and his reputation as the originator of all things wireless was not diminished. At news of his death in Rome on 20 July 1937 there were two minutes' radio silence in Britain and United States. In Italy there was a five-minute silence, ordered by Mussolini, who had courted Marconi and who was the first at his deathbed. Later the Italian leader would have a hideous fascist mausoleum built in the grounds of the Villa Griffone to which Marconi's body was removed from its original grave in Bologna.

Professor Edward Appleton, who had confirmed the existence of the ionosphere by sending a signal vertically upwards and getting it back again, wrote in the *Daily Mail* on 21 July of Marconi's 'great technical achievements', all of which 'can be traced to Marconi's almost obstinate belief that there is no limit to which wireless waves will not travel'. But there was a limit to Marconi's conception of wireless and it was left to others to discover the astonishing possibilities presented by Hertzian waves.

* * *

It is a wonderful story: the world's first radio broadcast picked up by just a handful of amateur enthusiasts, the boats of the United Fruit Company carrying bananas from South America up to the East Coast ports and US Navy ships equipped with wireless. These 'listeners' have a new kind of wireless receiver, which is capable of picking up speech and music that can be heard in a telephone receiver. Most of their messages are in Morse Code, but on Christmas Eve 1906 they are startled to be entertained by the spoken word. The Master of Ceremonies is a Canadian scientist and inventor, Reginald Fessenden, who has an experimental coastal station at Brant Rock, Massachusetts. He recalled the occasion a quarter of a century later:

> The program on Christmas Eve was as follows: first a short speech by me saying what we were trying to do, then some phonographic music . . . Handel's Largo. Then came a violin solo by me, being a composition by Gounod called Oh Holy Night, and ending with the words 'Adore and be Still' which I sang one verse of, in addition to playing the violin, though the singing, of course, was not very good. Then came the Bible text 'Glory to God in the highest and on earth peace to men of good will' and finally we wound up wishing them a Merry Christmas and saying that we proposed to broadcast again on New Year's Eve.

This was Fessenden reminiscing in a letter he wrote from his home in Bermuda on 29 January 1932. He said the New Year's Eve broadcast had gone ahead with a slightly different programme and that the transmissions from Brant Rock had been picked up locally and in the West Indies, 1,400 miles away. He was replying to a request from Westinghouse Electric about the broadcasts and suggested that verification would be found in the logs for those dates of the US Navy vessels and United Fruit Company ships. However, there is a great mystery about the broadcasts which has never been solved: despite intensive searches over a number of years nobody has ever found any contemporary confirmation that they took place. And yet Fessenden was no charlatan: he was a brilliant and prolific inventor with a high enough reputation near the end of his life not to make anything up: he died on 22 July less than six months after he wrote the letter. He was not claiming a first, though others have done that on his behalf, or claiming that his 1906 broadcasts were a breakthrough. That had happened earlier when he was undoubtedly the first in the world to send a telephone message by wireless.

Fessenden was ten years older than Marconi, the son of a Canadian Anglican minister born in 1866. He was precocious but also capricious: he excelled as a boy at school but did not stick at anything. When still a teenager he became a teacher in Canada and at the age of seventeen accepted the offer of a post as the principal and only staff member of the Whitney Institute in Bermuda. It seems he was restless and always unsure where he wanted to focus his talents and interests. He had a pleasant time in Bermuda, adopted by a large family of one boy and nine girls, one of whom, Helen Trott, he married. A classicist with an interest in mathematics, he became intrigued by electricity, which he read about in journals such as *Scientific American*.

Fessenden took his first step on the road to becoming an inventor when, in 1885, he quit Bermuda and sailed for New York in search of work with Thomas Edison, or anyone in the exciting world of electrical innovation who would give him a job. He banged on Edison's door several times before his persistence paid off and he was put to laying cables for the General Electrical Company. From there he moved on to work on chemistry, modifying the light bulbs then being manufactured, worked for other companies, and in 1892 left the experimental labs for academia to take up a post as Professor of Electrical Engineering at Purdue University in Indiana. It was here he first began to take an interest in electromagnetism and Hertz's experiments. The giant electrical firm of Westinghouse took notice and arranged for him to become a professor at the Western University, Pennsylvania (renamed in 1908 the University of Pittsburgh). From 1893 to 1900 he worked as an academic and was also an engineer with his own consulting firm. These were the years when Marconi made his historic advances in the practical use of Hertzian waves. Fessenden regarded Marconi's spark transmitters and coherer receivers as unnecessarily crude and he got a chance to challenge him when the US Weather Bureau accepted his proposal that he create a wireless network to be used for warnings of hurricanes and storms at sea. His first station was at Cobb Island in Maryland, just south of Washington, DC. It was here that he first experimented with a new wireless receiver that was not limited to the registering of dots and dashes. It was quite different from Marconi's coherer or his later magnetic detector. Fessenden called it a barreter, a device born of his knowledge of chemistry as well as electricity, and it was capable of detecting a continuous wireless signal carrying sound waves that could be heard in an earphone. He modified a spark transmitter so that it could transmit those waves at a high enough frequency to carry sound.

Cobb Island was remote, reached only by boat. Fessenden lived there with his wife Helen and young son. They were joined by a Cornell graduate, Alfred Thiessen, who had been assigned to the project by the Weather Bureau. It was Thiessen on 23 December 1900 who received the first, barely intelligible, wireless telephone call sent over about a mile between two wireless masts on Cobb Island. 'One, two, three, four, is it snowing where you are Mr Thiessen? If it is, telegraph back and let me know.' That was the breakthrough.

In the early 1900s there was a great deal of speculation about the possibility of the wireless telephone. By 1902 it was already the subject of a variety song with a line in the chorus 'She's calling you alone, by wireless telephone.' The joke, popular with telephone engineers in the 1970s, that if you did not answer your wireless telephone you must be dead was already in circulation in 1902. There was a great deal of anguish about the possible consequences: the fact that the boss could always call you back from work. Speculation about husbands calling from the train to say they would be late was rife. In fact the world of the mobile phone, which eventually became a reality, was imagined long before it became technically feasible. Thomas Edison, however, was a sceptic, reputedly dismissing Fessenden's notion of transmitting speech with: 'Fezzie, what do you say are man's chances of jumping over the moon? I think one is as likely as the other.'

Fessenden's major breakthrough was his discovery that a high-speed alternator could produce Hertzian waves. At the time he first experimented with this in the early 1900s it was still widely believed that only a spark could generate electromagnetic waves as a series of pulses. With Fessenden's alternator, built for him by the brilliant General Electric engineer, the Swede Ernst Alexanderson, it was possible to send signals that could carry speech and music. The alternators would also send Morse signals at a much greater speed than the spark transmitter and the Marconi Company soon acquired one.

Whether or not Fessenden's 1906–7 broadcasts took place, he had by that time demonstrated that wireless telephony worked. However, he was being funded by two wealthy businessmen who urged him to compete with Marconi on long-distance wireless telegraphy. The Brant Rock station was set up in 1905 with that intention and another station was established on the west coast of Scotland on the Mull of Kintyre, in Argyll. Some messages were sent and there was a claim that a telephone connection had been made. But the Scottish aerial blew down in a gale and was not repaired.

Fessenden's wireless telephony was soon outmoded when the radio valve was found to be capable of both transmitting and receiving speech. But radio telephony did not take off commercially. Instead the landline telephone became well established for use by individuals as well as companies and governments, while wireless telegraphy ruled the airwaves on land and on sea. It was this segregation of the phone network and the wireless network that had to be bridged for the cell-phone revolution to take place.

Radio, however, laid the cultural foundations. When equipment for transmitting and receiving speech and music first became available in kit form the wireless amateurs began to talk to each other and to create their own programmes. This was beginning to happen at the outbreak of the First World War and it flourished just after the war. It was the birth of broadcasting, first in America and then shortly after in Britain. Wireless telephony made headway on a much smaller scale, though the English satirist A.P. Herbert was predicting as early as 1920 that it would be a disaster. In *The Living Age*, on 7 August, he wrote on 'Modern Nuisances'.

> Wireless telegraphy was tedious enough; as it is, whenever one commits the tiniest murder one is hounded down and arrested by wireless in the S.S. *Argantic*. But wireless telephony seems to me to spell the end of civilisation. . . . The game keeper, the square leg umpire and every porter will have a set, even the bathing machine attendant. There will be no escape.

In that sense, the end of civilisation was more than half a century away. Instead of the wireless phone becoming ubiquitous it developed largely independently of the landline network. Fire brigades and police forces were quick to make use of the telephone and in both the United States and Britain police boxes appeared in the streets. These had a direct line to a police station and a handset that could be used by the public. Between the wars there were experiments in Germany and America with phones on railway trains but these never became commonplace. The 'walkie-talkie' first appeared in 1940, a two-way radio phone developed by Motorola in the United States for the Army.

The first mobile phone available to the public was the car phone introduced by Bell Telephones in St Louis, Missouri as early as 1946. For a long time it was thought that the only market would be for these cumbersome pieces of equipment. At first the price was prohibitive but as costs fell the demand rose and the airwaves became congested. Only a handful of car phones could operate in any one town and, as Richard Frenkiel pointed out, use of them was very frustrating. It was the Federal Communications Commission's decision to take away some of the wireless spectrum from US TV stations and hand it to the telephone companies that provided the stimulus to develop a system that could handle many more calls. Bell Telephone Engineers went back to a basic blueprint drawn up in 1947 and began to explore the possibility of the cellular phone. By the mid–1970s they had available the transistor, developed by the company's own engineers, and the microprocessor or minicomputer. Motorola added its own innovation with the first hand-held model, the DynaTAC, which would eventually be used in the first commercial American mobile network in Chicago in 1983.

By the early 1980s, Bell Telephones and Motorola, the trailblazers in the concept of a cellphone system, had been held up by slow-moving government action and had been overtaken by Japan and a consortium of Scandinavian countries calling themselves Nordic Mobile Telephone (NMT). It seems likely – though this part of the history of the mobile phone remains stubbornly obscure – that when the Japanese and NMT began to develop their cellular networks they borrowed the Bell blueprint. This seems to have been true of Britain, which adopted what Bell engineers called AMPS – the Advanced Mobile Phone System. Detailed accounts of its development were certainly freely available for anyone who wanted to read them.

It is a familiar story in the history of invention: first it is thought wireless telephony is impossible; then it is shown to be technically feasible but nobody is interested in it because existing technology appears to be more than adequate; then it is refined and demand begins to grow; and finally something that few thought anybody would want is in high demand.

AFTERWORD

THE NATURE OF INVENTION

A theme which has emerged from this study of the long gestation of five modern inventions is the importance of the outsider and amateur in making the breakthrough and achieving the 'eureka moment'. Their creations were inevitably crude and it would require all the investment and expertise of mainstream science and industry to turn the ugly duckling into a commercial swan. But time and again, even in the twentieth century, those innovations which have transformed our lives have been pioneered not by the big guns of established industries or the laboratories of the most brilliant scientists but by a few visionaries who had the temerity to imagine they could make the impossible possible. The question inevitably arises: why should this be? Is it just a quirk of history or is there some kind of logic to it?

Science can certainly inhibit innovation. The early aviators such as Sir George Cayley and Otto Lilienthal were liable to be ridiculed for their belief that manned flight was possible and few established scientists were prepared to damage their credibility by giving serious consideration to the idea. Octave Chanute, the distinguished engineer who gave the Wright brothers so much assistance, kept his interest in heavier-than-air flight under his hat until he retired. The Wright brothers did not fear ridicule, as they did not belong to a community of scientists and engineers. In this sense the amateur has an advantage over the professional.

In his pioneering attempts to 'see by electricity' John Logie Baird was generally regarded as a crank, a role he seems to have enjoyed much of the time. In the 1920s some scientists were interested in the possibility of transmitting pictures as well as words but believed, with some justification, that the technology was not available. No concerted effort to solve the problem was made until some years after Baird with his crude apparatus showed that television was feasible. Typically, the technology he put together was soon discarded and superseded by all-electronic television, which was created by scientists and engineers with much greater knowledge than the amateur Baird. But he was the one who first showed the world that television could be a reality.

The two key technologies which made the mobile phone possible, the telephone and wireless, were pioneered by amateurs: Alexander Graham Bell, who always styled himself a teacher of the deaf, and Guglielmo Marconi who told newspaper reporters he was 'an enthusiastic amateur of electricity'. The history of the mobile phone can be traced back at least as far as the Danish scientist Oersted, who first showed conclusively that the mysterious invisible forces of electricity and magnetism were related. Michael Faraday took that discovery and by ingenious experimentation produced the first electrical generator. Was Faraday an eminent scientist or an amateur? He was certainly self-taught, and the distinction between amateur and professional was not clear in the nineteenth century. James Clerk Maxwell who took Faraday's findings and devised a theory to describe the behaviour of electromagnetic wave, and Heinrich Hertz who conducted experiments to confirm Maxwell's predictions, were certainly both scientists. But they had no concept of wireless, leaving the practical application of their discoveries to the amateur Marconi. It is true that the scientist Oliver Lodge claimed to have sent messages electromagnetically before Marconi but this is generally discounted now.

The fact that scientists do not generally turn their discoveries into practical devices should not be surprising. It is not what they do. Their interest is focused on understanding, on teasing out the workings of the physical world. It is that understanding which makes innovation possible and nearly all inventors are ultimately absolutely reliant on the discoveries of science.

It is extremely unlikely that a scientist would have come up with the concept of the bar code. Joe Woodland, who first dreamed up the idea of adapting Morse Code for the identification of groceries, was a trained engineer, but there was nothing much in his technical education which suggested the innovation he proposed. More important was his time in the Boy Scouts, during which he was taught the dots and dashes of telegraph Morse. His attempt to build a working checkout system failed for lack of the appropriate technology. Science came to the rescue with the laser and the microchip, though those who made these high tech inventions had no idea they would be used at the supermarket checkout counter and, in time, for labelling almost everything.

That the amateur should have had a critical role to play in the creation of the personal computer is perhaps the most surprising discovery of this history. Not only did amateurs and outsiders play their part in the historical development of lithography and photography which proved crucial to the creation of the microchip, but the hackers and readers of *Popular Electronics* seized the opportunity to turn a chip intended for a calculator into the brain of a low-priced personal computer. Ed Roberts, who first saw the possibility of using Intel's integrated circuit to create a cheap computer, had a solid background in electronics and was not, in that sense, an amateur. But the company he had created in Albuquerque was tiny compared with the computer industry giants such as IBM.

The role of science in invention is to provide understanding and new materials with which the innovator can work. We should not expect scientists, however brilliant they are, to have the kind of practical turn of mind that dreams up new devices like the television or wireless. In the case histories presented here the only invention that was not dependent on previous scientific discovery was the Wright brothers'

Flyer. The study of aerodynamics was considered legitimate only for the understanding of ballistics and the heavier-than-air aviators of the early twentieth century had little scientific expertise to draw on. Wilbur and Orville Wright had to do all the calculations themselves before they could get off the ground. Their eminent rival Samuel Langley, whose powered planes were an embarrassing failure, clearly did not grasp even the need to understand aerodynamics.

If we should not expect scientists to come up with trailblazing inventions, what about the big guns of industry? The case of Alexander Graham Bell and Elisha Gray provides a classic example of the way in which entrenched interests resist the arrival of anything novel. Both Bell and Gray had backers who wanted them to devise a way of carrying multiple signals down a telegraph wire to solve a problem of congestion in the system. Both had the concept of the telephone and both were told by their backers to forget it. Gray was by far and away the more competent electrical engineer and he concluded that the telephone would probably be no more than a toy. Bell, the rank amateur, went ahead with the telephone, adding it as almost an afterthought to his patent for the multiple telegraph. He was the outsider who dared to propose something the telegraph industry regarded as a waste of time.

In the case of television, the giant Radio Corporation of America had no interest in it at all and waited for amateur developments to demonstrate its feasibility before putting their resources into it. In Britain, the British Broadcasting Corporation agreed to give television a trial run only when pestered by Baird. The BBC did nothing to develop it and neither did the manufacturers of wireless sets. Why should they use their resources to develop some new form of entertainment which, if it turned out to be popular, would compete with their established business, and, if a failure, would be a waste of money?

It was a mournful refrain of the supermarkets in the United States that none of the big electronics firms wanted to help them to find a way to computerise stock control and the checkout system. The grocery industry's profit margins were so small they did not look like potential customers for state-of-the-art electronic equipment. It was a campaign by the supermarkets to agree on a bar code system that finally galvanised the business machine industry into action. IBM, which had taken no interest in the project, made a last-minute bid to provide the equipment for the Universal Product Code and got the job.

Paul Eisler who devised the printed circuit in his lodgings in Hampstead just before the Second World War found that radio manufacturers did not want it because it meant making the girls who did the wiring in the factory redundant. Established businesses, Eisler found, were often opposed to innovation. Their own research teams might be put out if an amateur turned up with a bright idea they had not had themselves. It could always be said it would never work or that nobody would want it.

At the time the first home computers appeared there were several large established companies making computers for scientists and businesses. Why did none of them come up with this world-changing invention? It was left to Steve Wozniak, a self-confessed prankster and hacker, to take inspiration from the Altair devised by Ed Roberts and, with his friend Steve Jobs, devise the Apple 1. Wozniak learned about the Altair and the possibilities presented by the new integrated circuits at the Homebrew Computer Club in California. It was only when a demand for home computers became evident that the big companies moved in to start creating and selling their PCs.

When it comes to anticipating public demand it seems big business lacks the imagination to innovate. Even a firm such as Intel, which marketed some of the first microchips, was founded by a group of engineers and scientists who broke away from a large company to give themselves the freedom to try something new.

Eureka: How Invention Happens is itself intended as an innovative approach to the history of five inventions which play a large part in our lives. To put the spotlight on the amateur and outsider might appear to be overly romantic. But that emphasis was not sought at the outset: it was a surprising, and in some ways pleasing, discovery. The dotty individualist perhaps has a legitimate place in technological history: not the myth of the 'lone genius' revisited but respect for those whose dogged determination has sometimes defied the wisdom of the day and ushered in something magically new that few ever imagined could ever become a reality.

BIBLIOGRAPHY

Edgerton, David, *The Shock of the Old*, Profile Books, London, 2006

Johnson, Steven, *Where Good Ideas Come From: A Natural History of Innovation.* Allen Lane, London, 2010

1 The Bird Men

Ackroyd, J.A.D., *Sir George Cayley: The Invention of the Aeroplane near Scarborough at the Time of Trafalgar.* Journal of Aeronautical History paper no. 2011/6

Crouch, Tom D., *The Bishop's Boys: A Life of Wilbur and Orville Wright.* W.W. Norton, New York, 1989

Crouch, Tom D., 'Kill Devil Hills 17 December 1903', *Technology and Culture* 40, no. 3 (July 1999), pp. 594–8

Dee, Richard, *The Man Who Discovered Flight: George Caley and the First Airplane.* McClelland & Stewart, Toronto, 2007

Fairlie, Gerard B. and Elizabeth Cayley, *The Life of a Genius.* Hodder & Stoughton, London, 1995

Gibbs-Smith, Charles, *Sir George Cayley's Aeronautics, 1796–1855.* HMSO, London, 1962

— *Aviation: An Historical Survey from its Origins until the End of WW2.*. HMSO, London, 1970

— *The Rebirth of European Aviation 1902–1908.* HMSO, London, 1974

Harrison, Michael, *Airborne at Kitty Hawk: The Story of the First Heavier-than-air Flight made by the Wright Brothers December 17 1903.* Cassell, London, 1953

Hawkey, Arthur, *The Amazing Hiram Maxim: An Intimate Biography*, Spellmount Ltd, Staplehurst, Kent, 2001

Jakab, Peter L., *Visions of a Flying Machine: The Wright Brothers and the Process of Invention.* Airlife Publishing, Smithsonian Institution Press, Washington DC, 1990

Jarrett, Philip, *Another Icarus: Percy Pilcher and the Quest for Flight.* Smithsonian Institution Press, Washington DC 1987

Kelly, Fred C., *The Wright Brothers.* George Harrap, London, 1944

Kelly, Maurice, *Steam in the Air: The Application of Steam Power in Aviation during the Nineteenth and Twentieth Centuries.* Pen and Sword Aviation, Barnsley, 2006

Kirk, Stephen, *First in Flight: The Wright Brothers in North Carolina.* John F. Blair, Winston-Salem, North Carolina, 1995
Laurence, John, *Sir George Cayley: The Inventor of the Aeroplane.* M. Parrish, London, 1961
Lilienthal, Otto and Gustav, *Birdflight as the Basis of Aviation*, translated from the second edition by A.W. Isenthal. Longmans Green, London, 1911
McFarland, Marvin W. (ed.), *The Papers of Wilbur and Orville Wright, including the Chanute-Wright Letters and Other Papers of Octave Chanute*, 2 vols. McGraw-Hill, London, 2001
Padfield, G.D. and B. Lawrence, 'The Birth of Flight Control: An Engineering Analysis of the Wright Brothers' 1902 Glider', *Aeronautical Journal*, December 2003, pp. 697–718
Parramore, Thomas C., *First to Fly: North Carolina and the Beginnings of Aviation.* University of North Carolina Press, North Carolina, 2002
Penrose, Harald [*sic*], *An Ancient Air: A Biography of John Stringfellow of Chard, the Victorian Aeronautical Pioneer.* The Crowood Press Ltd, Marlborough Wiltshire, 2000
Root, Amos, 'The Wright Brothers' Flying Machine', *Gleanings in Bee Keeping*, January 1905
Runge, Manuela and Lukasch, Bernd, *Inventor Brothers: The Lives and Times of Otto and Gustav Lilienthal*, translated from the German by Jim Stearn. Berliner Taschenbuch Verlag, Berlin, 2007
Taylor, Bob, 'The Man Aviation History Almost Forgot: Charles E. Taylor', AvStop Magazine Online
Tobin, James, *First to Fly: The Unlikely Triumph of Wilbur and Orville Wright.* John Murray, London, 2003
United States Coastguard historian, 'The Indispensable Men', http://www.uscg.mil/history/articles/indispensable_men.asp
Vernon, 'The Flying Man: Otto Lilienthal's Flying Machine', *McClure's Magazine*, September 1894
Westcott, Lynanne and Degen, Paula, *Wind and Sand: The Story of the Wright Brothers at Kitty Hawk.* Abrams, New York, 1983
Wohl, Robert, *A Passion for Wings: Aviation and the Western Imagination 1908–1918.* Yale University Press, New Haven, Connecticut, 1994

2 Seeing with Electricity

Abramson, Albert, *Zworykin, Pioneer of Television.* University of Illinois Press, Champaign, Illinois, 1995
Baird, M., *Television and Me: The Memoirs of John Logie Baird.* Mercat Press, Edinburgh, 2004
— *Television Baird*, Haum, Cape Town, 1973
Barnard, G. P., *The Selenium Cell, its Properties and Applications.* Constable, London, 1930
Belrose, John S., *Fessenden's Christmas Eve Broadcast Revisited* http://www.radiocom.net/Fessenden/BelroseXmas.htm
Briggs, Asa, *The BBC: The First Fifty Years.* Oxford University Press, 1985
Brookman, Philip, Eadweard Muybridge. Tate, London, 2010
Burns, Russell W., *John Logie Baird: Television Pioneer.* Institute of Electrical Engineers, London, 2000
— *Television: An International History of the Formative Years.* IEE, London, 1998
— *Communications: An International History of the Formative Years.* IEE, London, 2004
Douglas, Susan J., *Inventing American Broadcasting 1899–1922.* Johns Hopkins University Press, Baltimore, Maryland, 1987
Farnsworth, Mrs Elma G. ('Pem'), *Distant Vision: Romance and Discovery on an Invisible Frontier.* Pemeberly Kent Publishers, Salt Lake City, 1989
Fisher, David E. and John Marshall, *Tube: The Invention of Television.* Counterpoint, Washington DC, 1996
Hendricks, Gordon, *Eadweard Muybridge: The Father of the Motion Picture.* Secker & Warburg, London, 1975
Hubbell, R. W., *Four Thousand Years of Television.* George Harrap, London, 1946

Jenkins, C. Francis, *Boyhood of an Inventor.* (Self published), Washington DC, 1931
— *Radiomovies, Radiovision, Television.* (Self published), Washington DC, 1929
— *Vision by Radio, Radio Photographs, Radio Photograms.* (Self published), Washington DC, 1925
Kamm A. and M. Baird, *John Logie Baird: A Life.* National Museums of Scotland Publishing, Edinburgh, 2002
Magoun, A. B., *Television: The Life Story of a Technology.* Greenwood, Westport, Connecticut, 2007
Mannoni, Laurent, *The Great Art of Light and Show: Archaeology of the Cinema.* University of Exeter Press, 2000
McArthur, T. and P. Waddell, *The Secret Life of John Logie Baird.* Hutchinson, London, 1986
McLean, Donald F., *Restoring Baird's Image.* IEE, London, 2000
Moseley, S. John, *Baird: The Romance and Tragedy of the Pioneer of Television.* Odhams Press, Burns, 1952
Muybridge, Eadward, *Animals in Motion* Chapman & Hall, London 1899
O'Hara, J. G. and W. Pricha, *Hertz and the Maxwellians: A Study and Documentation of the Discovery of Electromagnetic Wave Radiation, 1873–1894.* Peregrinus, London, 1987
Robinson, E. H. and G. Cock, *Televiewing.* Selwyn & Blount, London, 1935
Solnit, Rebecca, *Motion Studies: Eadweard Muybridge and the Technological Wild West.* Bloomsbury, London, 2003
Susskind, C., *Heinrich Hertz: A Short Life.* San Francisco Press, San Francisco, 1995
Wade, Nicholas, *A Natural History of Vision*, MIT Press, Cambridge, MA, 1998
— 'Philosophical Instruments and Toys: Optical Devices Extending the Art of Seeing', *Journal of the History of Neurosciences*, 13, no. 1 (2004), pp. 102–24
Webb, R.C., *Tele-visionaries: The People behind the Invention of Television.* Wiley for IEE, 2005

3 Written in the Sand

Beck, Francis Jr, *Some Reminiscing: A Few Years before the UPC.* http://idhistory.com/rca/FranBeckthoughtsbeforeUPC.pdf
Bertolotti, Mario, *The History of the Laser.* Institute of Physics Publishing, Bristol, 2005
Brown, Stephen A., *Revolution at the Checkout Counter: The Explosion of the Bar Code.* Harvard University Press, Harvard, 1997
Collins, David Jarrett, *KarTrak: The First Barcode Scanner.* Data Capture Institute, Duxbury, MA, 2010
Hecht, J., *Laser Pioneers.* Academic Press, Amsterdam, 1992
Laurer, George J., *Engineering was Fun: An Autobiography*, 3rd edn, Lulu Press, online publisher, 2012
Leibowitz, Ed, 'Bar Codes: Reading between the Lines', *Smithsonian*, 29 issue 11 (1999), pp. 130–46
Maiman, Theodore, *The Laser Odyssey.* Laser Press, Blaine, WA, 2000
Palmer, Roger C., *The Bar Code Book.* 3rd edn, Helmers Publishing, Peterborough, New Hampshire, 1995
Phaniteja, Janaswamy and P. Tom Derin, 'Evolution of the Bar Code', *International Journal for Development of Computer Science and Technology*, vol 1, issue 2
Seideman, Tony, *Barcodes Sweep the World.* www.barcoding.com
Townes, C. H., *How the Laser Happened: Adventures of a Scientist.* Oxford University Press, Oxford, 1999
Varchave, Nicholas, 'Scanning the Globe', *Fortune*, 149, no. 11 (31 May 2004)

4 Homebrewed

Agar, Jon, *The Government Machine: A Revolutionary History of the Computer.* MIT Press, Ann Arbor, 2003

Allen, Paul, *Idea Man: A Memoir of the Co-founder of Microsoft*. Portfolio/Penguin, London, 2011

Austrian, Geoffrey D., *Herman Hollerith, Forgotten Giant of Information Processing*. Columbia University Press, New York, 1982

Babbage, Charles, *Passages from the Life of a Philosopher*, edited with a new introduction by Martin Campell-Kelly. Pickering, London, 1994

Babbage, Henry Prevost, *Memoirs and Correspondence of Major General H.P. Babbage*. W. Clowes, London, 1910

Berlin, Leslie, *The Man Behind the Microchip: Robert Noyce and the Invention of Silicon Valley*. Oxford University Press, Oxford, 2005

Brock, David and David Laws, *The Early History of Microcircuitry: An Overview*. IEEE Annals of the History of Computing vol. 34, no 1, January–March 2012, pp. 7–19

Ceruzzi, Paul E., 'Ready or Not, Computers Are Coming to the People: Inventing the PC', *Organization of American Historians Magazine of History*, vol 24, issue 3 pp. 25–8, 2010

Collins, Graham P., 'Claude E. Shannon: Founder of Information Theory', *Scientific American*, 14 October 2002

Eisler, Paul, *My Life with the Printed Circuit*, edited with notes by Mari Williams. Associated University Presses, London and Toronto, 1989

Essinger, James, *Jacquard's Web: How a Hand-loom Led to the Birth of the Information Age*. Oxford University Press, Oxford, 2004

Freibert, Paul and Michael Swaine, *Fire in the Valley: The Making of the Personal Computer*. McGraw-Hill, New York, 2000

Gernsheim, Helmut, *The History of Photography vol. 2 The Rise of Photography 1850–1880*. Thames & Hudson, London, 1988

Gertner, Jon, *The Idea Factory: Bell Labs and the Great Age of American Innovation*. Penguin Press, London, 2012

Harrison, Graham, *The History Men: Helmut Gernsheim and Nicéphore Niépce*. Photo Histories 27 May 2014 http://www.photohistories.com/Photo-Histories/59/the-history-men-helmut-gernsheim-and-nicephore-niepce

Hobsbawm E.J. and Joan Wallach Scott, 'Political Shoemakers', *Past & Present*, 89, November 1980, pp. 86–114

Lathrop, Jay W., *The Diamond Ordnance: Fuze Laboratory's Photolithographic Approach to Microcircuits*. IEEE Annals of the History of Computing vol. 35, no 1, January–March 2013, pp. 48–55

Lawrence, Ashley, *A Message Brought to Paris by Pigeon Post 1870–71*. http://www.microscopy-uk.org.uk/mag/indexmag.html?http://www.microscopy-uk.org.uk/mag/artoct10/al-pigeonpost.html

Levy, Steven, *Hackers: Heroes of the Computer Revolution*, O'Reilly Media, Free Online Edition, 2010

MacHale, Desmond, *George Boole: His Life and Work*. Boole Press, Dublin, 1985

Mims III, Forrest M., *Siliconnections: Coming of Age in the Electronic Era*. McGraw-Hill, New York, 1986

— *The Altair Story: Early Days at MITS*, Creative Computing, 10 (November 1984)

Reid, T.R., *The Chip: How Two Americans Invented the Microchip and Launched a Revolution* Random House, London, 2001

Riordan, Michael and Lillian Hoddeson, *Crystal Fire: The Birth of the Information Age*. W.W. Norton, New York, 1997

Senefelder, Alois, *The Invention of Lithography*, translated from the German by J.W. Muller. The Fuchs & Lang Manufacturing Company, 1911 (Project Gutenberg)

Shurkin, Joel N., *Broken Genius: The Rise and fall of William Shockley, Creator of the Electronic Age*. Macmillan, New York, 2006

Swade, Doron, *The Cogwheel Brain: Charles Babbage and the Quest to Build the First Computer*. Little, Brown, New York, 2000

Turing, Alan, *On Computable Numbers, With an Application to the Entscheidungsproblem*, Proceedings of the London Mathematical Society, Series 2, 42 (1936–37), pp. 230–65
Twyman, Michael, *Breaking the Mould: The First Hundred Years of Lithography*. British Library, London, 2001
Watson, Roger and Helen Rappaport, *Capturing the Light*. Macmillan, London, 2013
Wolfe, Tom, 'The Tinkerings of Robert Noyce: How the Sun Rose on Silicon Valley', *Esquire*, December 1983, pp. 346–74
Wolff, Michael F., Interview with Robert N. Noyce for the Institute of Electrical and Electronics Engineers, Inc. IEEE Global History Network 19 September 1975
Wozniak, Steve with Gina Smith, *iWoz Computer Geek to Cult Icon: Getting to the Core of the Apple's Inventor* Headline Review, London, 2006

5 Hard Cell

Agar, John, *Constant Touch: A Global History of the Mobile Phone*. Icon Books, London, 2013
Aitken, Hugh G.J., *Syntony and Spark: The Origins of Radio*. John Wiley, Chichester, 1976
Belrose, John S., *Fessenden's Christmas Eve Broadcast Revisited* http://www.radiocom.net/Fessenden/BelroseXmas.htm
Bowers, Brian, *Sir Charles Wheatstone 1802–1875*. Institution of Electrical Engineers in association with the Science Museum, London, 2001
Bruce, Robert Bell, *Alexander Graham Bell and the Conquest of Solitude*. Gollancz, London, 1973
Christensen, Dan Charly, *Hans Christian Oersted: Reading Nature's Mind*. Oxford University Press, Oxford, 2013
Crowther, J.G., *Famous American Men of Science* [includes Joseph Henry]. Secker & Warburg, London, 1937
Engel, Joel S., 'The Early History of Cellular Telephone', *IEEE Communications Magazine*, August 2008, pp. 27–9
Frenkiel, Richard H., *Cellular Dreams and Cordless Nightmares: Life at Bell Laboratories in Interesting Times*. http://www.winlab.rutgers.edu/~frenkiel/dreams
Grosvenor, Edwin S. and Morgan Wesson, *Alexander Graham Bell: The Life and Times of the Man who Invented the Telephone*. Abrams, New York, 1997
Hamilton, James, *Faraday: The Life*. HarperCollins, London, 2002
Hong, Sungook, 'Marconi and the Maxwellians: The Origins of Wireless Telegraphy Revisited', *Technology and Culture*, 35, no. 4, October 1994
— *Wireless from Marconi's Black-box to the Audion*, MIT Press, Ann Arbor, 2001
Hounshell, David A., 'Elisha Gray and the Telephone: On the Disadvantages of Being an Expert', *Technology and Culture*, 16, no. 2 (April 1975), pp. 133–61
John, Richard R., *Network Nation: Inventing American Telecommunications*. Belknap Press of Harvard University Press, Harvard, 2010
Klemens, G., *Cellphone: The History and Technology of the Gadget that Changed the World*. McFarland, Jefferson, NC, 2010
Mahon, Basil, *The Man who Changed Everything: The Life of James Clerk Maxwell*. John Wiley, Chichester, 2003
Marconi, Degna, *My Father, Marconi*. F. Muller, London, 1962
Spufford, Francis, *Backroom Boys: The Secret Return of the British Boffin*. Faber and Faber, London, 2003
Tolstoy, Ivan, *James Clerk Maxwell: A Biography*. Canongate, Edinburgh, 1981
IT History Society http://ithistory.org/index.php

ACKNOWLEDGEMENTS

Three of the chapters in this book – those on the bar code, the mobile phone and the personal computer – have their 'eureka moment' in the mid–1970s, and the events described have barely made it into mainstream history. I was therefore reliant on the testimony of some of those who witnessed key events and who were generous with their help. David Collins provided material on the KarTrak bar code system on which he worked as well as some biographical details. My contact at the ID History Museum (http://www.idhistory.com), Bill Selmeier, provided much background information and put me in touch with a number of people involved in devising the UPC. Susan Woodland, daughter of the late Joe Woodland, sent me archival material and filled in some details of the story. George Laurer, who devised the bar code for the UPC, gave valuable background material. Judy Deeter of the Troy Historical Association tracked down some of those involved in the first UPC scan in 1974. Myra Borshoff Cook unearthed material from the Marsh Supermarkets archive.

A special thanks to Richard Frenkiel for his assistance with the history of the cell phone, drawing on his own experience and suggesting others I could contact about its development in the mid–1970s. Gerry diPiazza described the excitement of the early days and Thomas Haug provided background on the international picture. Martin Cooper, who commissioned the first Motorola mobile phone, the DynaTAC, was very generous with his time and assistance as I tried to get a better understanding of the birth of this new technology. Any research into the history of wireless leads inevitably to the wonderful website created by Thomas H. White: United States Early Radio History (http://earlyradiohistory.us).

Forrest Mims III outlined the pioneer days of the personal computer and was very generous with his time. Doron Swade answered a great many questions about Charles Babbage and his influence on the development of the computer. On the relationship between lithography and photography, I was fortunate to be able to call on the expertise of Roger Watson, curator of the Fox Talbot Museum, and Dr Michael Pritchard, Director General of the Royal Photographic Society. The only, and very brief, biography of John Benjamin Dancer, pioneer of microphotography, was kindly photocopied and sent to me by the Manchester Literary and Philosophical Society. Leslie Berlin, biographer of Robert Noyce, kindly answered some questions I had about his role in inventing the microprocessor.

One of the great puzzles that arose in my research into television was why it is we see a sequence of still images as movement, an effect historically referred to as 'the persistence of vision'. Emeritus Professor of Psychology at Dundee University, Nick Wade, generously provided much historical material and drew my attention to the importance of scientific 'toys' in the eighteenth and nineteenth centuries in studies of visual perception. Some additional research was undertaken by Thelma Rumsey. For material from the Lance Sieveking archive on his experience in very early television I am grateful to the Lilly Library at Indiana University. Some reminiscences of John Logie Baird were kindly forwarded by Gordon Sweet and John Heys of the Hastings Electronics and Radio Club.

John Ackroyd was generous with his expertise on Sir George Cayley and early experiments with flight and corrected some of the errors in my draft chapter. Bernd Lukasch of the Otto Lilienthal Museum in

Anklam, Germany, kindly emailed me a translation of the recent biography of the Lilienthal brothers, which was indispensable. Brian Riddle, Chief Librarian at the National Aerospace Library in Farnborough, drew my attention to the extensive archives on Cayley and to the wealth of visual material relating to the history of flight.

Though the internet provides more and more research material I remain grateful to the staff of both the London Library and the IET Library.

The concept of this book grew from a suggestion by Heather McCallum at Yale University Press and I am grateful to her for taking a close interest in its evolution and her very helpful observations on the draft text. I would like to thank Beth Humphries for her meticulous editing, and Rachael Lonsdale and Tami Halliday at Yale for their enthusiastic work gathering in the illustrations, making corrections, and getting the book into shape over several months. I thank Douglas Matthews for compiling an impeccable index. Any errors that remain are my responsibility. Finally I would like to thank Charles Walker at United Agents for looking after my interests as always.

INDEX

PICTURE CREDITS

1. Image courtesy of www.tvhistory.tv; 2. (National Aerospace Library)/ Mary Evans Picture Library; 3. (National Aerospace Library)/Mary Evans Picture Library; 4. Photo by Ottomar Anschütz, 1894. Archive Otto-Lilienthal-Museum/www.lilienthalmuseum. de; 5. Photo by Ottomar Anschütz, 1894. Archive Otto-Lilienthal-Museum/www. lilienthalmuseum.de; 6. Library of Congress (LC-DIG-ppprs-00626); 7. Library of Congress (LC-DIG-ppprs-00546); 8. Courtesy of Special Collections and Archives, Wright State University, Dayton, Ohio (40.3); 9. Courtesy of the Center for History of Sciences, The Royal Swedish Academy of Sciences; 10. Galleryhip; 11. © Illustrated London News Ltd/Mary Evans; 12. Mary Evans Picture Library; 13. © Photo Researchers/Mary Evans Picture Library; 14. ID History Museum; 15. Courtesy of HRL Laboratories, USA; 16. Courtesy of David Collins; 17. Courtesy of Ian Ormerod: NCR Fellowship UK; 18. Mary Evans/ Library of Congress; 19. Mary Evans Picture Library; 20. Library of Congress (LC-USZ62-118841); 21. DeGolyer Library, Southern Methodist University, Texas Instruments Records; 22. Intel Free Press; 23. Courtesy of the Computer History Museum; 24. Copyright © 1974 Poptronix, Inc; 25. Mac-History; 26. Courtesy of the Computer History Museum; 27. The Danish National Museum of Science and Technology;

28. Mary Evans Picture Library/Imagno; 29. © Photo Researchers/Mary Evans Picture Library; 30. © Photo Researchers/Mary Evans Picture Library; 31. Reprinted with permission of Alcatel-Lucent USA Inc; 32. General Electric installation manual, 1950s; 33. Dyna, LLC.